Mitigating Bias in Machine Learning

Mitigating Bias in Machine Learning

Carlotta A. Berry

Brandeis Hill Marshall

McGraw Hill books are available at special quantity discounts to use as premiums and sales promotions, or for use in corporate training programs. To contact a representative, please visit the Contact Us pages at www.mhprofessional.com.

Mitigating Bias in Machine Learning

1 2 3 4 5 LBC 28 27 26 25 24

Library of Congress Control Number: 2024942600

ISBN 978-1-264-92244-4
MHID 1-264-92244-2

Sponsoring Editor Lara Zoble	**Indexer** Edwin Durbin
Project Manager Tasneem Kauser, KnowledgeWorks Global Ltd.	**Production Supervisor** Richard Ruzycka
Acquisitions Coordinator Olivia Higgins	**Composition** KnowledgeWorks Global Ltd.
Copy Editor Lisa McCoy	**Illustration** KnowledgeWorks Global Ltd.
Proofreader Raghu Narayan	**Art Director, Cover** Anthony Landi

To my husband and daughter for sharing me with the world
—Carlotta A. Berry, PhD

To my husband, my parents, and my ancestors
—Brandeis Hill Marshall, PhD

We dedicate this work to the diverse voices in Artificial Intelligence (AI), who work tirelessly to call out bias and work to mitigate it and advocate for ethical AI every day. Some of the trailblazers doing the work are Ruha Benjamin, Timnit Gebru, Casey Fiesler, Joy Buolamwini, Rumman Chowdhury, and Safiya Noble. We also dedicate this work to the future engineers, scientists, and sociologists who will use it to inspire them to join the charge.

Contents

Preface xv
Acknowledgments xxix

1 Beyond Algorithmic Bias 1
  Learning Objectives 1
  Chapter Overview 1
  1.1 Introduction 2
  1.2 Beyond Ethics in AI 2
    1.2.1 Section Summary 2
  1.3 What Is Algorithmic Justice? 3
    1.3.1 The Main Causes of Injustice in Machine Learning 3
    1.3.2 The Sources of Harm in a Machine Learning
           Life Cycle 3
    1.3.3 Section Summary 5
  1.4 Definitions of Algorithmic Fairness 6
    1.4.1 Fairness Through Unawareness 6
    1.4.2 Individual Fairness 6
    1.4.3 Group Fairness 6
    1.4.4 Counterfactual Fairness 7
    1.4.5 Section Summary 7
  1.5 Fairness Metrics 7
    1.5.1 Statistical Parity Difference 7
    1.5.2 Equal Opportunity Difference 8
    1.5.3 Average Odds Difference 8
    1.5.4 Disparate Impact 8
    1.5.5 Theil Index 9
    1.5.6 Section Summary 9
  1.6 Methods for Fair Machine Learning 9
    1.6.1 Preprocessing 9
    1.6.2 In-processing 9
    1.6.3 Postprocessing 10
    1.6.4 Reweighing 10
    1.6.5 Adversarial Debiasing 10
    1.6.6 Reject Option-Based Classification 11
    1.6.7 Section Summary 11
  1.7 Tools to Help Detect and Mitigate Bias in Machine
       Learning Models 11
    1.7.1 AI Fairness 360 11
    1.7.2 Aequitas 11
    1.7.3 Section Summary 12

1.8   Best Practices to Build a Fairer Application   12
        1.8.1   Section Summary   13
1.9   Chapter Summary   13
Chapter Glossary   13
References   15
End of Chapter Questions   16

2   **Going Beyond the Technical: Exploring Ethical and
    Societal Implications of Machine Learning**   17
Learning Objectives   17
Chapter Overview   18
2.1   Introduction   18
2.2   Programming Approaches   18
2.3   Societal and Cultural Implications of Algorithms   19
        2.3.1   Racial Socialization   19
        2.3.2   Racism in Its Many Forms   19
        2.3.3   Why Does This Matter to Students?   20
2.4   Ethical Implications of Algorithms   21
2.5   Approaches to Mitigate Algorithmic Bias   23
2.6   Speak Up: Communicating Ideas with Digital Storytelling ...   24
        2.6.1   Digital Storytelling   24
        2.6.2   Other Considerations   25
2.7   Chapter Summary   26
Chapter Glossary   26
References   27
End of Chapter Activities   29
        Reflective Exercises   29
        Algorithmic Bias and Search Algorithms   30
        Digital Storytelling Extensions   31
        Get Involved and Take Action!   31

3   **Social Media and Health Information Dissemination**   33
Learning Objectives   33
Chapter Overview   34
3.1   Introduction   34
        3.1.1   Mitigating Bias in Health Information
                on Social Media   34
3.2   MyHealthImpactNetwork: For Students by Students   35
        3.2.1   Why Consider Social Media?   36
        3.2.2   Studying MyHealthImpact Twitter Feed   36
        3.2.3   Data Preprocessing   36
        3.2.4   Check Missing Values   37
        3.2.5   Check Multicollinearity   38
3.3   Results of Data Inferential Analysis   39
        3.3.1   Time Series Analysis   39
3.4   Chapter Summary   44

Chapter Glossary 45
References 45
End of Chapter Questions 47

4 Comparative Case Study of Fairness Toolkits **49**
   Learning Objectives 49
   Chapter Overview 49
   4.1 Introduction 50
   4.2 Bias 50
         4.2.1 Label Bias 50
         4.2.2 Sampling Bias 51
         4.2.3 Representation Bias 51
   4.3 Fairness 51
         4.3.1 Statistical Measures 53
         4.3.2 Similarity Measures 53
   4.4 Applying Responsible AI 53
         4.4.1 Checklists 53
         4.4.2 Software Toolkits 53
         4.4.3 Fairness Software Toolkits 54
         4.4.4 Evaluating Fairness Toolkits 55
   4.5 Results 56
   4.6 What Are the Limitations of These Toolkits? 58
         4.6.1 Diverse Range of Biases 58
         4.6.2 Diverse Measures of Fairness 58
         4.6.3 Bias Detection 59
         4.6.4 Bias Mitigation 59
         4.6.5 Intersectional Analysis 59
         4.6.6 Applicable to Data Without Sensitive Information 59
   4.7 Chapter Summary 59
   Acknowledgments 59
   Chapter Glossary 60
   References 61
   End of Chapter Questions 62

5 Bias Mitigation in Hate Speech Detection **71**
   Learning Objectives 71
   Chapter Overview 71
   5.1 Introduction 71
   5.2 Background 72
         5.2.1 Case Study of Hate Speech Detection 73
         5.2.2 Section Summary 74
   5.3 Bias in Hate Speech Detection Systems 75
   5.4 Bias Mitigation in Hate Speech Detection
         Using Transfer Learning 75
         5.4.1 Case Study of Transfer Learning 76
         5.4.2 Section Summary 77

5.5   Bias Mitigation in Hate Speech Detection
      Using Transfer Learning   77
      5.5.1   Case Study on Multitask Learning   78
      5.5.2   Section Summary   79
5.6   Adversarial Methods for Bias Reduction
      in Hate Speech Detection   79
      5.6.1   Case Study on Adversarial Training   80
      5.6.2   Section Summary   81
5.7   Benefits and Pitfalls   81
      5.7.1   Transfer Learning   81
      5.7.2   Multitask Learning   81
      5.7.3   Adversarial Methods   82
      5.7.4   Section Summary   82
5.8   Other Methods   82
      5.8.1   Section Summary   83
5.9   Hands-on Exercise   83
      5.9.1   Model   83
      5.9.2   Dataset   83
      5.9.3   Data Preprocessing   83
      5.9.4   Hate Speech Classification   84
      5.9.5   Evaluation   85
      5.9.6   Bias Visualization   85
5.10 Chapter Summary   85
Chapter Glossary   86
References   87
Further Reading   88
End of Chapter Problems and Questions   88
Problems   89

6 **Unveiling Unintended Systematic Biases in Natural
   Language Processing** **91**
Learning Objectives   91
Chapter Overview   92
6.1   Introduction   92
      6.1.1   Pause Giant AI Experiment   92
      6.1.2   Why Do We Trust AI?   93
      6.1.3   Why Are There So Many Challenges?   93
6.2   Unfairness and Bias in NLP Applications   94
      6.2.1   Recycling the Same Biases   94
      6.2.2   AI Incident Repositories   94
6.3   Bias Taxonomy   95
      6.3.1   Denied Opportunities and Preconceived Views   96
      6.3.2   Biases Are Everywhere   97
6.4   Mitigating NLP Bias and Unfairness   98
      6.4.1   Find and Neutralize   98
      6.4.2   Measure and Evaluate   98
      6.4.3   Examine Biases   99

6.5   Chapter Summary 99
Chapter Glossary 100
References 100
End of Chapter Questions 103

7   **Combating Bias in Large Language Models** **105**
Learning Objectives 105
Chapter Overview 106
    Prerequisites 106
7.1   Introduction 106
    7.1.1   Bad Data In, Bad Data Out 106
7.2   Vectorization of Stochastic Parrots 107
    7.2.1   Linear Analogies 108
    7.2.2   Section Summary 110
7.3   Natural Language Processing: Linear Decision Making
      for Nonlinear Language 111
    7.3.1   Attention Layer Mathematics 112
    7.3.2   Section Summary: Outcome of the Attention
            Weights 113
7.4   Stage One: Data Collection 114
    7.4.1   Dataset Nutrition Labels 114
    7.4.2   Data Cards 115
    7.4.3   Data Documentation 115
7.5   Stage Two: Fight Bad Math with Better Math 115
    7.5.1   Counterfactuals 116
    7.5.2   Parity 116
    7.5.3   Stratified Sampling 117
7.6   Stage Three: Model Constraints/Operations 117
    7.6.1   Flagging 117
    7.6.2   Pruning 118
    7.6.3   Nudging 118
Case Study: The Limits of Better Training Data with
No Constraint 119
7.7   Chapter Summary 120
Chapter Glossary 120
References 122
End of Chapter Questions 122

8   **Recognizing Bias in Medical Machine Learning and AI Models** ... **125**
Learning Objectives 125
Chapter Overview 125
8.1   Introduction 126
8.2   Defining Machine Learning 126
    8.2.1   Supervised Machine Learning 127
    8.2.2   Unsupervised Machine Learning 127
8.3   Building a Simple Machine Learning Model: Use Case 1 127
    8.3.1   Preparing Our Development Environment 127

8.3.2   Downloading Anaconda by Going to
www.anaconda.com/download   127
8.3.3   Updating Anaconda Packages from Terminals   128
8.3.4   Installing the Prospector Static Code Analysis Tool   128
8.3.5   Installing jupyter-matplotlib Visualization
Library: ipympl   128
8.3.6   Standard Steps to Build a Machine Learning Model   128
8.3.7   Classifying Digits with the k-Nearest Neighbor
Algorithm   129
8.4   Health Care Bias and Inequities: Use Case 2   129
8.4.1   Background   129
8.4.2   Are We There Yet?   129
8.4.3   The Use Case   130
8.4.4   Implementation Results and Brief Explanation   131
8.5   Chapter Summary   134
Chapter Glossary   134
References   134
End of Chapter Questions   136

9   **Toward Rectification of Machine Learning Bias in Health Care
Diagnostics: A Case Study of Detecting Skin Cancer
Across Diverse Ethnic Groups** **137**
Learning Objectives   137
Chapter Overview   138
9.1   Introduction   138
9.1.1   How Does ML Bias Occur, and How Do
We Mitigate It?   138
9.2   Case Study: Mitigating Bias in ML for Melanoma   139
9.2.1   Retraining Melanoma Detection Algorithms
for Diverse Skin Tones   142
9.2.2   The Diverse Dermatology Images Dataset   143
9.3   Defining Types of Biases and Mitigation Techniques
in ML Life Cycles   145
9.4   Machine Learning Fairness   149
9.5   Chapter Summary   150
Chapter Glossary   151
References   152
End of Chapter Questions   155

10   **Applying the Wells-DuBois Protocol for Achieving Systemic
Equity in Socioecological Systems** **157**
Learning Objectives   157
Chapter Overview   158
10.1   Introduction   158
10.1.1   Understanding AI and ML Use in
Socioecological Systems   158
10.1.2   Examples of Socioecological Inequity and Bias   159

        10.1.3 Clustering in ML for Model Outcome Assessment 159
        10.1.4 Basic Concepts and Definitions 160
   10.2 Equity Framework and Tool Application 160
        10.2.1 The Systemic Equity Framework 160
        10.2.2 The Wells-DuBois Protocol 161
        10.2.3 Similarities and Differences with Other
                Equity Tools 161
        10.2.4 Section Summary 162
   10.3 Clustering Overview and Application 162
        10.3.1 A Socioecological Example on Food Spending 163
        10.3.2 Visualization and Initial Analysis 164
        10.3.3 Section Summary 167
   10.4 Applying the Wells-DuBois Protocol 167
        10.4.1 Section Overview 169
   10.5 Discussion and Future Directions 169
        10.5.1 Other Clustering Activities to Help Achieve
                Systemic Equity 169
        10.5.2 Further Discussion on Socioecological Systems 169
        10.5.3 Other Benefits of Employing the Wells-DuBois
                Protocol and Systemic Equity 170
        10.5.4 Section Summary 170
   10.6 Chapter Summary 170
   Chapter Glossary 171
   References 171
   End of Chapter Problems 174

11 **Community Engagement for Machine Learning** **177**
   Learning Objectives 177
   Chapter Overview 178
   11.1 Introduction: Principles and Components
         of Community Engagement 178
        11.1.1 Prerequisite Knowledge and Context: Case Study
                of Flint, Michigan 178
        11.1.2 Brief Introduction to Environmental Justice
                and Environmental Data Justice 180
        11.1.3 Data Justice 182
        11.1.4 What Is a Community? 183
        11.1.5 Community-Based Participatory Research 184
        11.1.6 Citizen Science and Community Science 184
        11.1.7 Section Summary 186
   11.2 Project Initiation: Steps of Conducting Community-Driven
         Environmental Data Science 186
        11.2.1 Phase 1: Building Partnerships 188
        11.2.2 Phase 2: Preparation for Modeling: Done
                in Conversation with Stakeholders 190
        11.2.3 Phase 3: Model Development 197
        11.2.4 Phase 4: After Modeling 197

11.3   How to Engage Communities in the Process: Case Study
       of the Mobile Lead Testing Unit Project in
       Newark, New Jersey   198
       11.3.1   Brief History Newark Lead Crises   198
       11.3.2   Project Initiation: The Newark Water Coalition
                and the Initiation of the Mobile Lead Testing Unit   199
       11.3.3   Identifying Stakeholders   201
       11.3.4   Data Collection   201
       11.3.5   Building Community Capacity   202
11.4   Chapter Summary   202
Chapter Glossary   203
References   203
End of Chapter Problems   206

Preface

Motivation

We grew up in the pre-Google times where we needed to go to the physical library, sift through library catalogs, and peruse the stacks in order to *hopefully* find the relevant information we needed. If the book was not there, then we could ask the librarian when it would be returned. Sometimes the librarian told us who checked it out and we could harass our fellow friend/classmate to borrow the book. Fast-forward to now where we're living our lives simultaneously in the physical and digital world. The information that we used to find only in books and talking with others is now readily available on the Internet. We now rely on this digital infrastructure to perform everyday tasks like buying groceries, paying bills, and providing us with real-time directions.

However, while the opportunity for access to information has expanded drastically, the quality of that information and its impact on other systems, tools, and platforms—both in our physical and digital worlds—are still being rooted out. We have experienced a dilution of our physical privacy with increased digital surveillance in our neighborhoods with doorbell cameras, in shopping centers, and at traffic intersections. We have become more aware of the higher likelihood of the digital misclassification of Black and Brown faces, which has led to the misplaced criminalization of innocent people. We have become accustomed to living and learning within the limits of our digital infrastructure, and for some of us, these limits are more severe.

So, here is where this book, *Mitigating Bias in Machine Learning*, enters the landscape. The instructional content, especially in the natural sciences, have unfortunately overlooked the cascading tensions and frictions induced by our expanding digital space. We set out to contribute to the educational resources available that specifically discusses the impact of all these digital products and suggest interventions to lessen their harmful effects.

Artificial intelligence (AI), a term first coined in the 1950s, was originally defined as "the science and engineering of making machines intelligent" by John McCarthy. By the 1990s, artificial intelligence's description had become more concrete and evolved to "the designing and building of intelligent agents that receive percepts from the environment and take actions that affect that environment" in a foundational AI textbook, *Artificial Intelligence: A Modern Approach*." These definitions, predating the Internet, were established by white, male computer science faculty at top-ranked U.S. institutions. The premise was that much of the foundational AI work was designed and developed by this demographic group. But with the proliferation of the Internet, social networks, and apps, AI moved from an abstract concept at an academic setting to being accepted as a decision-making agent affecting millions of people around the world.

As AI development progressed, the description reflected a much more relatable and controversial stance as programs with the ability to learn and reason like people. And a key subdivision, machine learning, emerged to specify the advanced computations and optimizations needed toward attempts to achieve the learning and reasoning goals of AI. Soon after AI was introduced, machine learning (ML) emerged in the late 1950s. ML approaches are heralded as a set of computing algorithms with the ability to learn without being directly programmed. ML's original goal was to record or remember all the positions it had seen before and ultimately outsmart the algorithm's programmer.

We use AI throughout our daily lives such as with geospatial positioning systems (GPSs). Being able to navigate a new city in real time is made possible by AI systems ingesting the thousands of roadways connecting cities and countries. We are the beneficiaries of ML approaches when it comes to product recommendations. We can more easily find comparable products, compare and contrast their reviews, and add supplemental products to our e-commerce shopping carts.

With these benefits, there is a growing list of instances where AI and ML produced inaccurate outcomes. A predominant misguided AI example comes in the mis-identification of Black and Brown people in open criminal cases, which have led to their unfortunate detainment from hours to days. Several of these individuals are now suing the police department due to the injustice. Using ML methods, one bad outcome is the lower credit card limits for U.S. women and the higher number of credit cards collected by women. Credit lending to U.S. women is less than 50 years old and was decided in the 1974 Equal Credit Opportunity Act (ECOA). It granted U.S. women the right to open credit cards in their own name. Compounded with pay inequity, U.S. women receive these lower credit card limits, so they tend to have more credit cards than their male counterparts. The historical and socioeconomic contexts that influence computational decisions and outcomes are why we set out to create this book. We wanted to highlight examples of disparities and pinpoint recommendations to reduce the harms generated by computationally based systems. Recently, there have been many cases of bias in AI that have garnered national media attention. For example, chatbots that become racist or bigoted over time, algorithms that convert highly pixelated images to white people, hiring apps that filter out women applicants, facial recognition technology that does not recognize black faces, AI generating all non-Black or non-Brown professor images, or recidivism technology that targets one demographic of perpetrators. All these challenges motivate the need for diverse perspectives in AI such as amplifying the voices of women, Black, Latino, Hispanic, Indigenous, disabled, LGBTQ+, and international scholars. This text has answered the call and will address that need.

Introduction

This textbook was a long time in the making but a much-needed addition to the burgeoning landscape of ML. Although ML has recently experienced exponential growth, there are a few texts about the presence of bias. As presented in *Data Conscience: Algorithmic Siege on Our Humanity* (Marshall, 2022):

"bias has become the ethical catchall in tech/data. Bias has two distinct definitions that makes the choice of this word a bit unclear. Understanding bias takes on one or these two extremes: combating prejudice or correcting mathematically based errors. Thus, the goals for addressing bias in tech systems do not overlap. Conflating these meanings here keeps the tech community stuck in defining bias as correcting errors, rather than mitigating prejudice and errors. These harms persist perpetuating chaos and confusion in, around, and outside of tech spaces. Bias, even under the umbrella of these two definitions, takes on many, many forms (for instance, https://catalogofbias.org/biases, which does not contain an exhaustive list)."

We therefore tackle bias directly to promote discussions and tangible methods that integrate ethics in robotics, AI, natural language processing, and ML.

This textbook is ideal for undergraduate or graduate students or those seeking an introduction to ML. Since there are few textbooks with practical applications of ML, this contribution will fill in the gap by introducing the topic with an emphasis on a real-world perspective and implementations.

Book Organization

The book is organized to tell the story of how bias in ML has negatively impacted certain communities. It will highlight ML approaches and how bias in these digital structures has failed specific communities, including historically excluded, marginalized, and minoritized communities, then how these failures have led to injustice for these populations. Each chapter is designed to make the reader think critically about the systems and platforms and how the oppression and inequities have been scaled, in addition to providing recommendations that may serve as a plan of action to mitigate the biases identified there.

The layout of this book takes the reader on a journey that starts with how disparities show up in various contexts. Then it highlights applications and toolkits that identify bias in text-based systems. Then we move on to health care and how bias can impact the care and treatment of vulnerable populations. Finally, environmental justice in ML addresses everyday unavoidable but hidden quality-of-life issues.

Each chapter will start with a question related to some field of ML that will be addressed. It will then highlight the learning objectives that should be achieved in the journey to answer the motivating question. Next, will be an overview that may include highlighting the key terms discussed or defined in the chapter. The chapter will present the answer to the question in the context of practical applications that may include examples, user or case studies, experiments, or simulations. Finally, there will be a summary of how the learning objectives were achieved as well as recommendations for promoting ethics and/or mitigating bias in the relevant field. The end of chapter questions, problems, and activities will enable the reader to appreciate the richness of the field and understand the theory through active learning. We hope you learn and enjoy.

Chapter Themes

The following list summarizes the chapters organized by major themes.

Bias and Ethics in Machine Learning

- "Beyond Algorithmic Bias" by Ana Carolina da Hora and Silvandro Pereira Pedrozo answers the question, "What is algorithmic justice and why do we need to talk about it?"
- Brooke Odle, Katherine Finley, Victoria Longfield, and Rodrigo Serrão provide context for the text in their chapter on the question: "What does it mean for a machine learning algorithm to be ethical?"
- "Where and how can we incorporate different perspectives in the design of machine learning?" is answered by Bavisha Kalyan, Anthony Diaz, and Mara Carrasquillo.

Large Language Learning Models and Bias

- Jazmia Henry explores the question: "How does bias show up in large language models and how can we combat that bias?" in "Combating Bias in Large Language Models."

Bias in Frameworks/Fairness in Systems

- In "Comparative Case Study of Fairness Toolkits," Keith McNamara, Jr., Kiana Alikhademi, Brianna Richardson, Emma Drobina, and Juan E. Gilbert answer the question: "How effective are fairness toolkits at detecting and mitigating bias and ensuring fairness in machine learning? And what are their shortcomings?"

Bias in Frameworks/Fairness in Software

- "How can you mitigate bias in hate speech detection systems?" is the question Zahraa Al Sahli answers in "Bias Mitigation in Hate Speech Detection."

Health Care

- In "Toward Rectification of Machine Learning Bias in Health Care Diagnostics: A Case Study of Detecting Skin Cancer Across Diverse Ethnic Groups," Jennafer Roberts and Laura Montoya answer the question: "How can we mitigate bias in health care in machine learning?"
- Isaac K. Gang asks "How can you mitigate bias in medical machine learning and AI systems?" in the chapter "Recognizing Bias in Medical Machine Learning and AI Models."
- A concern in the health care field is "Why does undertheorizing magnify ethical issues in health AI/ML among diverse populations?," and this will be explored by Fay Cobb Payton, Xuan Lui, and Lynette Yarger.

Socioecological Systems

- Ayushi Aggarwal, Tyrek Shepard, Thema Monroe-White, and Joe F. Bozeman III explore the question: "What overarching tools and concepts should you use when attempting to yield equitable outcomes in social and ecological (socioecological) systems?" in "Applying the Wells-DuBois Protocol for Achieving Systemic Equity in Socioecological Systems."

Natural Language Processing

- "How do systemic biases emerge in natural language processing, what societal impacts do they create, and how can we address these biases?" is answered by Olga Scrivner in "Unveiling Unintended Systematic Biases in Natural Language Processing."

Biographies

Ayushi Aggarwal is a staff professional at Geosyntec Consultants who works on remediation projects, phase 1 site assessments, and environmental process consultancy. She graduated with a master's in environmental engineering from Georgia Institute of Technology, Georgia, USA, and bachelor's in chemical engineering from the Vellore Institute of Technology, India. As an undergraduate and graduate researcher, her contributions range from nutrient recovery to data analysis and prediction using neural networks. Her master's thesis explored an inequity checklist–based system's usage to detect systematic biases in a public food-energy system using data visualization tools like clustering and expanded the study concerning chemical exposure to humans.

Dr. Kiana Alikhademi is a senior machine learning engineer at Walmart Global Tech. In 2023, Dr. Alikhademi received her PhD from the Department of Computer and Information Science and Engineering at the University of Florida. She also holds a master's degree in computer science from the same institution. Dr. Alikhademi earned her bachelor's degree in computer science from the Amirkabir University of Technology in Tehran, Iran.

Zahraa Al Sahili is a PhD candidate at Queen Mary University of London and holds the prestigious DeepMind PhD Scholarship. Her research primarily focuses on vision language foundation models, with a keen interest in fairness and robustness in AI. She received her master's in electrical and computer engineering degree from the American University of Beirut with machine learning as a research area, emphasizing transfer learning, computer vision, and graph neural networks. Al Sahili has also been involved in academic teaching, including teaching assistant roles, curriculum developer, and as an instructor for UCL, QMUL, Udacity, AUB, and University of Groningen. Furthermore, Zahraa Al Sahili has been featured on "Tech Sisters Stories," a platform that shares stories of women in tech.

Dr. Carlotta A. Berry is a professor and endowed chair in electrical and computer engineering at Rose-Hulman Institute of Technology. Her area of expertise is human-robot interaction, engineering, and mobile robotics education. She is the author of multiple technical publications and one textbook, *Mobile Robotics for Multidisciplinary Study*. She is a prolific speaker, mentor, role model, and STEM trailblazer. In her efforts to increase the number of women and historically marginalized and minoritized students earning degrees in computer science and computer, electrical, and software engineering at her university, she co-founded the Rose Building Undergraduate Diversity professional development, networking, and scholarship program. She also worked with other faculty to create the first multidisciplinary minor in robotics at Rose-Hulman. In 2020, to achieve her mission to diversify STEM by bringing robotics to people and bringing people to robotics, she launched her business, NoireSTEMinist educational consulting. She also co-founded Black In Engineering and Black In Robotics to promote diversity, equity, inclusion, and justice in STEM. Her novel strategies to normalize seeing Black women in STEM, including performing robot hip hop slam poetry, writing black STEM romance novels, conducting robotics workshops, creating open source robots, sharing Black STEM digital AI art, and using social media to educate the world about engineering and robotics, have proven to be groundbreaking. Due to her innovative work in engineering education and STEM outreach, she has appeared in several print and digital media including *Forbes, Black Enterprise, New York Times*, and *CBS News*. She has been recognized with several national awards, including the American Society of Engineering Education (ASEE) fellow, ASEE Electrical and Computer Engineering Division Distinguished Engineering Educator, Grace Hopper Celebration Educational Innovation Abie Award, Institute of Electrical and Electronic Engineers Undergraduate Teaching Award, Indiana Business Journal Women of Influence, and Society of Women Engineers Distinguished Engineering Educator.

Dr. Joe F. Bozeman III is an assistant professor of civil and environmental engineering with a courtesy appointment in the School of Public Policy at the Georgia Institute of Technology. His research focuses on developing equitable circularity, urban carbon management, and food-energy-water strategies. He has over a decade of private and public sector experience, and his award-winning work has been featured in major media outlets such as *Popular Science*, the *Geographical Magazine*, and NPR. He really enjoys transdisciplinary collaboration and believes approaching research questions in this way is a must for addressing the complex, "wicked" challenges of our time. On a personal level, he also enjoys sound engineering (various genres), multimedia production (including for science communication), watching and participating in sports/athletics, and video game playing when time permits (role-playing and sports games).

Ana Carolina da Hora is a master's student in ethics in computer vision at Unicamp (2023–2025) and has a bachelor of computer science degree from the Pontifical Catholic University of Rio de Janeiro, with diverse experiences throughout the undergraduate program. She graduated from the Apple Developer Academy program at the same university, where she experienced the development of applications with machine learning. She was president of the IEEE chapter of the university between 2016 and 2020 and worked for two years at the Cyberlabs Research Laboratory, focusing on image and voice recognition research. She serves as a columnist on artificial intelligence and ethics at MIT Technology Review Brazil and UOL Tilt, and has also written for *Canal Futura, Gizmodo*, and *Folha de São Paulo*. Since 2021, she has been part of the research group at the Center for Technology and Society at the FGV Rio Law School, working on cybersecurity and algorithmic justice themes. In 2020, she was elected one of the 100 important women researchers in the world in the field of ethics in AI. In 2021, she was listed in the Forbes Under 30 by MIT. In 2022, she also received the Sabia Award from the Cambridge Education Department for her research published during her undergraduate studies, titled "Algorithmic Racism in Facial Recognition." In 2022, she was recognized by the Secretary of Women of Rio de Janeiro as one of the female leaders in technology in the city, through the Nilse da Silveira Award. She is a consultant for the United Nations, Superior Electoral Court, UNDP, and presidency of the Brazilian government, where she is part of the council for social, sustainable, and economic development. She is also a science communicator in the projects Computação sem Cão, funded by Serrapilheira from 2018 to 2020, and Ogunhe Podcast. In 2020, she founded the Instituto da Hora, a research institute focused on digital rights in Brazil.

Anthony Diaz is the co-founder and executive director of the Newark Water Coalition. He has traveled nationally and internationally to work on water issues on indigenous lands and rural communities. Anthony was able to represent Newark at the United Nations Climate Conference in Glasgow, Scotland. He believes that people power can win against the interconnected struggles of humanity. The Newark Water Coalition is a multiethnic and multigenerational grassroots organization that provides services as well as resources to the Newark community. The Newark Water Coalition has been serving the community since 2018. In 2023, the Newark Water Coalition distributed over 150,000 pounds of food and over 100,000 gallons of water, serving over 20,000 community members annually. Anthony, in partnership with Bavisha Kalyan, a doctoral student at the University of California Berkeley, initiated the Mobile Lead Testing Unit, a community science project based in Newark. This collaboration aimed to bridge the gap between quantitative science and community engagement, fostering a unique model for inclusivity and shared decision making. This project allowed the Newark Water Coalition to take on additional lead research projects with the Stevens Institute of Technology and Rutgers University School of Public Health. The Newark Water Coalition, alongside Bavisha Kalyan, published a community voice manuscript in a special edition of the *Environmental Justice Journal* titled "Community Scientists of the Newark Water Coalition Are a New Dawn for Community-Owned and Managed Research Projects: Mobile Lead Initiative."

Emma Drobina is a PhD student in human-centered computing at the University of Florida, studying under Dr. Juan Gilbert as part of the Computing for Social Good Lab. She previously received a master's in computer science from the University of Florida and a bachelor's of science in computer science from the University of South Carolina. Her research focuses on explainable and interpretable machine learning.

Dr. Katherine Finley is an assistant professor of philosophy at Hope College (PhD, University of Notre Dame). She works primarily on topics at the intersection of philosophy of mind and cognitive science and applied ethics—specifically in the ethics of technology and bioethics. Much of her work is highly interdisciplinary—she has collaborated with colleagues in engineering, computer science, psychology, neuroscience, and anthropology on various research and teaching projects. She has recently published on topics including ethical issues in engineering computing; ethical and therapeutic issues arising from the use of artificial intelligence in digital phenotyping in psychiatry; embodied cognition and the impact of computational metaphors for the mind and brain; cognitive penetration and the perception of race; and issues of hermeneutic injustice in relation to mental disorders—and in venues including *Philosophical Psychology, Ergo, The American Journal of Bioethics*, and in volumes from Routledge, MIT Press, Brill, and Wiley-Blackwell. She has also disseminated her work through podcast and YouTube interviews and public presentations through organizations like the National Alliance on Mental Illness (NAMI). Additionally, she is currently running two studies—one investigating how memories of moral or immoral motivations for actions are impacted by "in-group" and "out-group" biases, specifically in relation to race, gender, and political affiliation, and another, using transcranial magnetic stimulation (TMS) to investigate the effects of firsthand narratives on stigma and empathy toward those who experience psychosis and the underlying neural mechanisms involved in these processes. She is also passionate about teaching and has collaborated with colleagues in engineering and neuroscience to create ethics modules for their courses, and is committed to engaging students in philosophy (especially ethics) outside the typical university walls and has taught courses in multiple prison education programs as well as community organizations (e.g., homeless shelters, senior education organizations).

Dr. Isaac K. Gang is an associate professor in the College of Engineering and Computing, MS Data Analytics Program at George Mason University. He joined the CEC faculty in the fall of 2020 from Texas A&M University-Commerce where he served as an assistant professor of computer science as well as the department's outreach coordinator. Before coming to TAMUC, Gang was an assistant professor of computer science and engineering at the University of Mary Hardin-Baylor and an adjunct professor of computer science at the University of Southern Mississippi's School of Computing before joining UMHB. His current and primary teaching responsibilities at Mason largely involve data analytics graduate capstones and a mix of computer science and applied information technology courses. Gang is a former DOE grant winner, president, and board member of the Association of Computer Educators in Texas (ACET), Industry Advisory Board (IAB) coordinator, and the director of CS for All. His primary research agenda involves big data/analytics, cybersecurity (ransomware and steganography), and image/signal processing.

Dr. Juan E. Gilbert is the Banks Family Preeminence Endowed Professor and chair of the Computer and Information Science and Engineering Department at the University of Florida where he leads the Computing for Social Good Lab. Dr. Gilbert is a fellow of the ACM, IEEE, the American Association of the Advancement of Science (AAAS), and the National Academy of Inventors. In 2012, Dr. Gilbert received the Presidential Award for Excellence in Science, Mathematics, and Engineering Mentoring from President Barack Obama. In 2023, Dr. Gilbert was named a laureate of the National Medal of Technology and Innovation by President Joe Biden for pioneering and championing universal design in elections technology to make voting accessible for everyone and increasing diversity in the computer science workforce. Dr. Gilbert received his MS and PhD degrees in computer science from the University of Cincinnati and his BS in systems analysis from Miami University in Ohio.

Jazmia Henry is a highly skilled Data Leader with a wealth of experience in Machine Learning, Data Science, and Large Language Models. Currently pursuing a DPhil in Social Data Science at the University of Oxford, Jazmia is actively engaged in cutting-edge research with a focus on Generative AI. This part-time academic pursuit complements her role as a Founder and CEO of Iso AI where she contributes to state-of-the-art Reinforcement Learning and Generative AI solutions and her former role as a Senior Data Scientist at Microsoft.

With a decade of experience, Jazmia's expertise spans A/B Testing, Experimentation, and Thought Leadership. Her work during her tenure as a Stanford CSRE Practitioner Fellow and Affiliate Fellow at Stanford HAI reflect her commitment to diversity and inclusion. She has spoken at conferences across North America, including at notable events such as the Black Women in Data Summit and NeurIPS. Furthermore, her role as the Head of Machine Learning at Motley Fool from 2020 to 2022 saw her build a Machine Learning team with strong AI Ethical principles.

Jazmia's career journey also includes impactful roles at Morgan Stanley, where she served as a Data Strategist, Ward Black Law as a Marketing Analyst and Hillary for America.

In addition to her corporate roles, Jazmia has actively contributed to academia as a Lead Instructor for Correlation One's Data Science program and as a Reviewer and Area Chair for the Neural Information Processing Systems conference. Her current research focuses on what she calls the "mechanisms of action" of AI- an exploration into the components of AI reasoning that lead to outcomes that affect the general public. You can find more information about her work on her website jazmiahenry.com.

Bavisha Kalyan has been a doctoral student at University of California, Berkeley since 2018, with a focus on environmental engineering. Her doctoral work, on measuring and predicting lead hazards, led her to building a machine learning model to predict lead hotspots. In pursuit of a more nuanced and locally contextualized approach to machine learning, Bavisha created a partnership with Anthony Diaz, the executive director of the Newark Water Coalition. Together, they initiated the Mobile Lead Testing Unit, a community science project based in Newark. This collaboration aimed to bridge the gap between quantitative science and community engagement, fostering a unique model for inclusivity and shared decision-making. With the Newark Water Coalition, Bavisha published a community-voice manuscript titled "Community Scientists of the Newark Water Coalition Are a New Dawn for Community-Owned and Managed Research Projects: Mobile Lead Initiative" in the journal *Environmental Justice Journal*. This publication explains the process of creating the Mobile Lead Testing Unit and serves to demonstrate importance of community-academic partnerships. The insights gained through the Newark Water Coalition project are the inspiration for Bavisha's chapter. Bavisha is driven by the hope that her experiences will encourage other quantitative-focused scientists to step outside their comfort zones. She emphasizes that scientists should work with integrity to ensure their scientific work not only advances knowledge but also actively protects and enhances the well-being of the communities it engages with.

Dr. Xuan Liu obtained her PhD in statistics from North Carolina State University advised by Dr. Daowen Zhang. She has broad research interests, including but not limited to missing data, categorical data analysis, generalized linear models, and statistical testing. Her doctoral dissertation topic is "Generalized Score Test on Categorical Data with General Missing Data Patterns'". She graduated from the University of Minnesota Twin Cities with a dual degree in mathematics (BA) and statistics (BS), where she found her interests in combinatorics and graph theory. She was advised by Dr. Victor Reiner on the topic "Hurwitz Actions and Orbits of Coxeter Elements in Complex Reflection Groups" during her undergraduate work. She also worked with Dr. Travis Scrimshaw and presented their work at the Nebraska Conference for Undergraduate Women in Mathematics (NCUWM) 2018 and published their work on Annales Henri Poincaré. Dr. Liu has a solid mathematical and statistical background and rich experiences in data analysis. Besides her dissertation work, she also collaborated with Duke CFAR and UNC SESH Global for AIDS Research during her PhD program and published a peer-reviewed journal article on clinical infectious diseases as a co-author. She is currently pursuing her industry career at Google as a data scientist.

Dr. Victoria Longfield is the associate professor of digital liberal arts at Hope College where she works within the Center for Teaching and Learning to support digital humanities and digital technology initiatives within faculty and student scholarship and in classroom settings. Longfield is the inaugural digital liberal arts specialist of the thriving digital liberal arts program at Hope College, where she has worked since 2016 to develop a program that supports over 80 classroom projects a year and countless new emerging technology initiatives across the entirety of the campus. She holds her master's of science in library and information science from the University of Illinois and is currently ABD for her PhD in information studies from Dominican University. For the past decade, she has been researching and writing about the integration of digital competency skills into the undergraduate classroom and is currently conducting a study to develop a digital competency skill framework for undergraduate teaching for her PhD work. Longfield is passionate about teaching undergraduate students digital skills that are foundational for their life in a global and digital society and supporting faculty in their work to educate students in this innovative way.

Dr. Brandeis Marshall is founder and CEO of DataedX Group, a data ethics learning and development agency that helps managers and practitioners integrate equity into their data operations and practices. Trained as a computer scientist and as a former college professor, she teaches, speaks, and writes about the racial, gender, socioeconomic, and socio-technical impact of data operations on technology and society. She wrote *Data Conscience: Algorithmic Siege on Our Humanity* (Wiley, 2022) as a counter-argument reference for tech's "move fast and break things" philosophy. She pinpoints, guides, and recommends paths to moving slower and building more responsible, human-centered AI approaches. She centers her work on making data and AI concepts snackable to understand for practical implementation from the classroom to the boardroom. As co-lead of the Atlanta Interdisciplinary AI Network, she's developing data citizens through humanities-centered critical data literacy community workshops and supporting new AI researchers. Also she provides data equity scholarship, professional development, and resources as a team member on the NSF Institute for Trustworthy AI in Law & Society, a partnership between the University of Maryland, George Washington University, and Morgan State University. Her thought leadership has appeared in *Heinrich-Böll-Stiftung, Medium, OneZero, The Moguldom Nation,* and on CNN. She has spoken to audiences across the AI and justice sectors including ACLU, Harvard, Kapor Center, Stanford, Truist, Urban League, and Visa.

Dr. Keith McNamara, Jr. graduated from the University of Florida with a doctorate in human-centered computing in 2023. Dr. McNamara received a bachelor's degree in computer science from the University of Maryland, Baltimore County in 2018.

Dr. Thema Monroe-White is an assistant professor in the Department of Technology, Entrepreneurship, and Data Analytics at Berry College and serves as the academic director of the Campbell Center for Data Analytics. As an interdisciplinary scholar, her work explores the systemic biases that affect the workforce and educational journeys of racially minoritized groups within science, engineering, and information technology fields. Her research concerns understanding the innovative pathways for achieving social and economic justice for minoritized groups via data, algorithmic and AI literacy, STEM education, and entrepreneurship. She holds a PhD in science, technology, and innovation policy from the Georgia Institute of Technology and master's and bachelor's degrees in psychology from Howard University.

Laura Naomi Montoya is the founder and managing partner of Accel Impact Ventures, the executive director of the Accel AI Institute, and the president of LatinX in AI (LXAI) organization. Her academic background is in biology, physical science, and human development. Laura has served as a director with Women Who Code, an advisor for

Udacity's AI and Data Nano degree, and an affiliate researcher with the Berkman Klein Center for Internet and Society at Harvard Law. Laura chairs and serves on program committees for research workshops at AI and ML conferences including NeurIPS, ICLR, ICML, and ACM FAccT. Recent research areas include reducing biased data representations in machine learning models, the effects of artificial intelligence development for developing countries, and paralleling biological and synthetic neural networks seen in mycology, entomology, and computational science. Laura has led sessions on social impact, tech diversity, and ethical AI development for Creative Mornings, Katapult Future Fest, Silicon Valley Future Forum, Tech Inclusion Conference, Thrival Summit, and Global Hive Summit and keynoted the "Future of Work" for the Data and Society Conference at UC Berkeley. Laura has given guest lectures and technical workshops at Google, Santa Clara University Law, Stanford University Computational Social Science, and GTC Deep Learning School. Recently she spoke at TEDx Santa Barbara and has been featured in WITtalks and CIIS podcasts, Xconomy, Verizon News, and Forbes Leadership.

Dr. Brooke Odle is an assistant professor in the Engineering Department at Hope College. She and her team of undergraduate researchers are interested in developing interventions to reduce the risk of musculoskeletal injuries in nursing students and personnel. Courses she teaches include "Engineering Computing," "Biomechanical Systems," "Bioinstrumentation Laboratory," "Dynamic Systems Laboratory," and "Mechanics of Materials Laboratory." With her colleagues in philosophy, sociology and social work, and digital liberal arts, she co-created a module for "Engineering Computing" to engage students in current events concerning the ethics and societal implications of computing, technology, and design. This work has been presented at American Society for Engineering Education (ASEE) conferences and is documented in conference papers, which can be found in the ASEE Papers on Engineering Education Repository. Prior to joining Hope College, Dr. Odle was a postdoctoral fellow in the Department of Biomedical Engineering at Case Western Reserve University. There, she worked in the Motion Study Laboratory at the Advanced Platform Technology Center within the Louis Stokes Cleveland Veterans Affairs Medical Center. She developed and evaluated control systems to restore standing balance after paralysis, explored experimental biomechanical and computational modeling techniques to investigate interactions between the upper extremities and walkers during static and quasi-static standing postures, and investigated the feasibility of neural stimulation to facilitate assisted transfers after paralysis. She received her PhD and MS in biomedical engineering from the New Jersey Institute of Technology. She received her BS in bioengineering from the University of Pittsburgh.

Dr. Fay Cobb Payton is a professor emerita and was a full professor (with tenure) of information technology/analytics at North Carolina State University. She is a named University Faculty Scholar for her leadership in turning research into solutions for society's most pressing issues. She is a visiting scholar and special advisor on inclusive innovation at Rutgers University. She completed a rotation as a program director at the National Science Foundation (NSF). She initiated the CISE Minority Serving Institution Research Expansion Program and worked on several initiatives, such as Smart Health and Biomedical Research in the Era of Artificial Intelligence and Advanced Data Science; AI Fairness, Equity, Accountability & Transparency Dear Colleague Letter; and Research Expansion and Cloud Computing Dear Colleague Letter with partnerships with

Amazon, Google, and Microsoft. She received the NSF Director's Award in 2020 for her work with NSF INCLUDES. Dr. Payton is a member of the National Academies of Science, Engineering and Medicine's Committee on Women in Science, Engineering and Medicine. She led the NSF funding and support of the National Academies of Science, Engineering and Medicine Consensus Study on Transforming Trajectories: Women of Color in Tech. She has published over 150 peer-reviewed journal articles, conference publications, and book chapters. She has appeared in or been cited by *Scientific American, MIT Tech Review, Essence Magazine, IEEE Spectrum, Insider Higher Ed, CBS Radio Network, Sunrise America, Financial Review,* and others. She earned a PhD in information and decision systems (with a specialty in health systems) from Case Western Reserve University. Prior to joining academe, she worked in engineering and consulting at IBM, Time, Inc., and EY. She is the author of *Leveraging Intersectionality: Seeing and Not Seeing* and serves on several advisories and boards.

Silvandro Pereira Pedrozo is a Brazilian expert in software quality assurance and machine learning, with a background in telecommunication and software engineering. He graduated from the University of Pernambuco with a bachelor's degree in telecommunication engineering and from CESAR School with a master's degree in software engineering. He worked for more than four years as a quality assurance software engineer at the Samsung Research and Development Institute, where he led a technical team and a software testing research group. He is currently a senior quality assurance software engineer at Encora, a global innovation services company. His master's thesis explored how to detect and reduce algorithmic bias in systems that use supervised machine learning for decision making. He has shared his knowledge and insights on algorithmic bias at college conferences. He is also a vocal online supporter of the need for constant verification and validation of systems that use artificial intelligence and impact the lives of people every day.

Jennafer Shae Roberts has been an associate researcher at the Accel AI Institute since 2017, specializing in ethics and artificial intelligence. With a master's in anthropology and social change, she brings a multidisciplinary approach to her work, focusing on correcting biases and exploitation around AI, data, and machine learning. Jennafer's publications include "Contextualizing Artificially Intelligent Morality: A Meta-Ethnography of Theoretical, Political, and Applied Ethics," which was presented at the Future of Information and Communications Conference, and "In Consideration of Indigenous Data Sovereignty: Data Mining as a Colonial Practice," which was presented at the Future of Technology Conference in 2023. Her work has been recognized by the Montreal AI Ethics Institute, demonstrating its impact. Collaboration is central to Jennafer's work. She co-authored *The Glamorization of Unpaid Labour: AI and Its Influencers* along with Nana Nwachukwu and Laura Montoya. Jennafer's passion for addressing ethical challenges in AI drives her contributions to the field.

Dr. Olga Scrivner is an assistant professor in computer science and software engineering at Rose-Hulman Institute of Technology. She joined the CSSE faculty in 2021 from Indiana University where she served as a research scientist at the Cyberinfrastructure for Network Science Center. Her research focuses on AI biases and skills gaps in the workforce and education. Scrivner is a board member at the American Council on Education for Women's Network Indiana and a Google Women Tech Makers Ambassador. She also serves on the external advisory board for the NSF-funded Advanced Cyberinfrastructure Coordination Ecosystem (ACCESS).

Dr. Rodrigo Serrão is a Brazilian sociologist living and working in the United States. He is an assistant professor of sociology at Hope College, located in Holland, Michigan. His research interests include race and ethnicity, religion, and immigration, with a focus on the Latinx experience in the United States. He has published several articles in academic journals and edited volumes in publications such as *Review of Religious Research, Latin American Perspectives, Perspectives of Religious Studies, International Journal of Latin American Religions, Sociology of Religion*, and *Religião & Sociedade* (Brazil). Currently, Dr. Serrão is engaged in a research project titled "Beyond Ethnicity: Perceptions of Racial Identity and Belonging Among Latinx Students." This project explores how Latinx students perceive their racial identity in relation to privilege, racism, colorism, and their sense of belonging within the context of a historically white college. The study also examines how these perceptions may differ based on the students' skin color and place of origin. Dr. Serrão's other project focuses on understanding how northeastern Brazilians, a historically discriminated-against group within Brazil, navigate both domestic and international migration amidst the deep political and ideological divisions in both Brazil and the United States. He holds a PhD in sociology from the University of South Florida and a graduate certificate in Latin American and Caribbean studies from the same university.

Tyrek Shepard is a researcher and health care technologist with a passion for leveraging digital solutions to improve health, wellness, and health care delivery. He earned his master's of science in public policy from the Georgia Institute of Technology and a bachelor's of science in accounting from North Carolina State University. While in graduate school, Tyrek completed his research training under the guidance of Dr. Thema Monroe-White by supporting her NSF-funded study on the challenges and experiences of underrepresented women of color in STEM innovation and entrepreneurship. Notable projects outside of this grant include his development of a novel application, Boddy Buddy, that leveraged personal health informatics to encourage healthy and sustainable weight gain, a master's-level environmental policy assessment and capstone for the South River Watershed Alliance, and a doctoral-level manuscript evaluating the policy narratives pertaining to the advancement and regression of transgender rights in the United States. Following his time in graduate school, Tyrek has served in research and product management roles in the pharmaceutical industry. His current experience is focused on user experience research for accessibility and service design at Bristol Myers Squibb in Princeton, New Jersey.

Dr. Lynette Yarger is a professor in the College of Information Sciences and Technology at the Pennsylvania State University. She earned her MS and PhD in computer information systems from Georgia State University's Robinson College of Business. Prior to joining the faculty at Penn State, Dr. Yarger worked as a software engineer at AT&T, a member of technical staff at Lucent Technologies, and a product life cycle manager at Avaya Communications.

Acknowledgments

This book has been years in the making. Others have tried, and failed, to capture the breadth and depth of disparities, inequities, and biases within the AI and machine learning space. We hope that we've been able to provide an insightful and thought-provoking discussion starter with our diversity of topics. We'd like to thank so many who have contributed to this book becoming a reality. It wouldn't have happened without you.

First, we'd like to give a special thank you to the chapter authors: Ana Carolina da Hora, Silvandro Pereira Pedrozo, Brooke Odle, Katherine Finley, Victoria Longfield, Rodrigo Serrão, Fay Cobb Payton, Keith McNamara, Jr., Kiana Alikhademi, Brianna Richardson, Emma Drobina, Juan E. Gilbert, Zahraa Al Sahili, Olga Scrivner, Jazmia Henry, Isaac K. Gang, Jennafer Shae Roberts, Laura Naomi Montoya, Ayushi Aggarwal, Tyrek Shepard, Thema Monroe-White, Joe F. Bozeman III, Bavisha Kalyan, Anthony Diaz, and Maya Carrasquillo. Your scholarship is a blend of science-based expertise and justice-centered advocacy. It's inspiring and will continue to encourage all who engage with your work.

To the book chapter reviewers, we'd like to thank you for sharing your perspectives and feedback with us: Kiana Alikhademi, Kenya Andrews, Junior Bernadin, Sarah Brown, Richard Charles, Tanner Durant, Lauren Maffeo, Dan Mantz, Anthony Ndolo, Jennafer Shae Robert, Mary Salami, Olga Scrivner, Tarek Shraibati, Shahram Najam Syed, Kush Varshney, Louvere Walker-Hannon, and Angela R. Wells.

To everyone at McGraw Hill, thank you for giving this much-needed book a platform to reach those in classrooms to the boardrooms.

Finally, we'd like to thank those of you who read, share, and comment on this book. We hope this book helps you critique AI and machine learning in new ways and motivates you to continue charting a path toward a more just digital future.

CHAPTER 1

Beyond Algorithmic Bias

Ana Carolina da Hora
Unicamp

Silvandro Pereira Pedrozo
Cesar School

Question: What is algorithmic justice and why do we need to talk about it?

Learning Objectives

Upon completion of this chapter, the student should be able to

- Understand the importance of ethics in artificial intelligence
- Learn about the main causes of injustice in machine learning
- Learn the different sources of harm in a machine learning life cycle
- Understand different definitions of algorithmic fairness
- Explore the concepts of fairness metrics and methods that we can employ to construct a machine learning model with greater fairness

Chapter Overview

In this chapter, we transcend the conventional discussions surrounding ethics in artificial intelligence (AI) and delve deeper into the domain of algorithmic justice. To ensure a comprehensive understanding, we analyze different perspectives and interpretations of algorithmic fairness, recognizing the diverse contexts in which fairness is applied. In pursuit of fairness, we then shift our focus to the practical aspects and explore a range of fairness metrics, and then we proceed to examine methods for fostering fair machine learning. This chapter serves as a resource for students and developers seeking to move beyond ethical and technical considerations and actively contribute to a future of AI that embodies fairness and inclusivity.

1.1   Introduction

Technological bias, or non-neutrality, is a classic philosophical and sociological issue. Lately, it has become a technology issue and, more recently, a technology problem. This raises question about the influence of technology on culture and society. Some advocates claim that certain technologies are neutral enough to make decisions for humans. However, these new technologies, such as facial recognition and textual predictive models, are not as neutral as some would have us believe. They are technologies created from the remnants of colonization based on prejudices and deficiencies that reinforce unequal worldviews. As a result, small groups insist on making technology as neutral as possible so as not to be held responsible for human decision making. The reflection of this lack of responsibility has been studied in algorithmic bias, which in its direct definition, occurs when AI makes unfair decisions for certain groups of people. This chapter explores strategies for mitigating algorithmic bias and reducing some negative consequences when AI technology is misused.

1.2   Beyond Ethics in AI

The complexity of human relationships is a subject of examination within the realm of ethics. With the progression of technological mediation in human interactions, the field of AI ethics has flourished. This field is dedicated to comprehending the mechanisms through which this mediation occurs and the significance of this vital area of research. This chapter is mainly aimed at addressing the ethical concerns that people may have about the design and application of AI systems. To formally define, at the heart of ethical AI, the idea is that it should never lead to wrong actions, the result of bad learning, or bad choice of datasets, which can impact safety and human dignity.

The problem is that AI is not necessarily ethical, and even ethical AI is not necessarily trustworthy. We need to go beyond this binary of thought. It is really important to deeply understand the problem that this ethical AI is trying to solve and also the data used to feed this model. Most of these AI systems are under the control of large companies and governments, among others, without auditing the algorithms and the development process and without transparency of the real uses of these creations. In Brazil, the use of facial recognition in public security has raised several ethical discussions about the perception of justice from automated decision making. For example, in 2021, 99 percent of those arrested unfairly for the use of these technologies were young Black people.

1.2.1   Section Summary

This section delved into the intricate nature of human relationships and its connection to ethics, particularly in the context of advancing technological interventions. It addressed the ethical apprehensions surrounding the design and implementation of AI systems. Additionally, the text emphasized that AI systems aren't inherently ethical, and the presence of ethics doesn't necessarily guarantee trustworthiness. This section advocates for moving beyond this binary perspective.

1.3   What Is Algorithmic Justice?

According to Verma and Rubin (2018), in recent years, the term "algorithmic justice" has increasingly attracted researchers' attention in AI, software engineering, and law. More than 20 different notions of fairness have been proposed by the scientific community in general. Even so, as it is a very sensitive and important issue, there is still no clear consensus on which definition applies in each different situation. According to Kleinberg et al. (2016), diverse researchers have shown that it is not possible to satisfy all definitions of fairness at the same time. Considering this scenario, it becomes extremely important to understand in depth the problem that a machine learning model is solving. Combine this understanding with an in-depth analysis of the database used for training and from there apply the definitions of justice that are aligned with the purpose of the proposed solution.

More and more often, machine learning algorithms are inserted into the daily lives of the population, whether making purchase suggestions, recommending TV shows, assisting in hiring employees, or making high-risk decisions on financial loan applications. Unlike humans, who need to rest, machines can work all the time and take into account a much more significant number of factors when making a decision. However, people have different biases that end up contributing to "unfair" decisions, and machines incorporate these biases into their development and dissemination. Bringing the concept of justice into the context of decision making, it can be said that equity is the absence of any prejudice or favoritism toward an individual or group based on their intrinsic or acquired characteristics.

1.3.1   The Main Causes of Injustice in Machine Learning

According to the work of Barocas and Selbst (2016), the five main causes of injustice in the area of machine learning are the following:

1. **Distorted sample**: Once some initial bias occurs, this bias can worsen over time.

2. **Contaminated examples**: Data labels are biased due to data labeling activities performed by biased humans.

3. **Limited features**: Features may be less informative or unreliably collected, causing a model error when building the connection between features and labels.

4. **Sample size disparity**: If the minority group and the majority group data are unbalanced, likely, the model is not good for the minority group.

5. **Representatives**: Some features represent sensitive attributes (e.g., neighborhood) and can cause a bias for the machine learning model even if sensitive attributes are excluded.

1.3.2   The Sources of Harm in a Machine Learning Life Cycle

According to Suresh and Guttag (2021), there are different sources of harm in a machine learning life cycle. These kinds of harms were listed as seven different types of bias. Each of these takes place in a different phase of the machine learning pipeline. Figure 1.1 shows the process of data generation, with the possible bias that may occur.

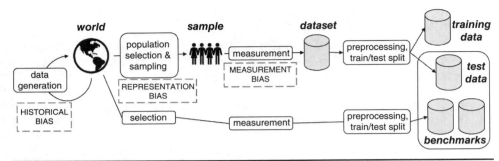

FIGURE 1.1   Data generation.

Source: Own authorship. Adapted from Suresh and Guttag (2021).

1. **Historical bias**: This type of bias occurs during the data generation process. Historical bias arises when there is a misalignment between the world as it is and the values or goals to be codified and propagated in a model. It is a normative concern with the state of the world, and it exists even with perfect sampling and feature selection.

2. **Representation bias**: This bias is present during the step of population selection and sampling. Representation bias occurs when algorithms are trained on data that are not representative of the population; it is linked to the problem with companies training facial recognition technologies. These technologies are trained primarily on pictures of white men, and for the most part with Western individuals.

3. **Measurement bias**: Afterward, the characteristics and labels present in this sample are identified and measured. Measurement bias occurs when the attributes that will be used are defined and measured and the characteristics chosen may not actually represent what you want to measure.

The four remaining types of bias take place during the model building and implementation phase of a machine learning model. Figure 1.2 shows this process and when each bias appears.

4. **Learning bias**: During the process of model optimization, the training data are fundamental. Learning bias can occur during this model learning step when modeling choices amplify performance disparities across different examples in the data.

5. **Aggregation bias**: It can occur during the machine learning model definition. Aggregation bias arises when a one-size-fits-all model is used for data where there are underlying groups or types of examples that should be considered differently. This bias can lead to a model that is not optimal for any group or a model that is adequate only for the dominant population.

6. **Evaluation bias**: After the model is defined and trained, it is important to evaluate it using test data and benchmarks. Evaluation bias may occur if the benchmark data used for a particular task do not represent the user population. Using a benchmark that is not representative will encourage the development and deployment of models that perform well only on the subgroup represented by the benchmark data.

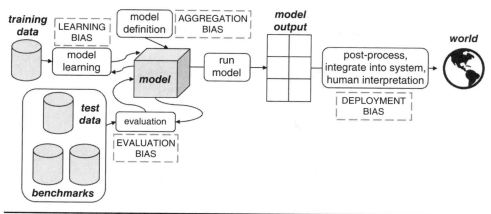

FIGURE 1.2   Model building and implementation.

Source: Own authorship. Adapted from Suresh and Guttag (2021).

7. **Deployment bias**: After the final model is integrated into the real world, it is important to keep monitoring the application, or deployment bias may occur. Deployment bias arises when there is a mismatch between the problem a model is intended to solve and how it is actually used. In some cases, systems produce results that must first be interpreted by human decision makers. Despite good performance alone, they can end up causing harmful consequences because of automation or confirmation bias.

According to Bellamy et al. (2018), a favorable label is a label whose value corresponds to a result that provides an advantage to the recipient. We can exemplify these advantages as receiving a loan, getting a job, or not being arrested. A protected attribute is an attribute that divides a population into groups that have parity in terms of benefits received. We can exemplify these groups as race, gender, social class, and religion. Protected attributes are not universal, but application specific. A privileged value of a protected attribute indicates a group that historically has a systematic advantage.

Considering that bias is a systematic error, in the context of algorithmic fairness, we are concerned about unwanted bias that puts privileged groups at a systematic advantage and nonprivileged groups at a systematic disadvantage. In this way, we can say that an algorithmic fairness metric is a quantification of undesired biases in data or training models, and a bias mitigation algorithm is a procedure to reduce undesired bias in these data or training models (Bellamy et al., 2018).

1.3.3   Section Summary

Within this section, the focus was on algorithmic justice, delving into its core concepts. The discussion revolved around five primary drivers of injustice and the seven kinds of bias within the realm of machine learning. Additionally, an exploration of potential origins of harm throughout the life cycle of machine learning processes was undertaken.

1.4   Definitions of Algorithmic Fairness

Research in the area of fairness has advanced and brought up a series of different fairness criteria in recent years. These criteria seek to operationalize certain concepts of equality on a mathematical level. Taking this into account, it is important to distinguish the main different concepts of fairness: fairness through unawareness, individual fairness, group fairness, and counterfactual fairness.

To help exemplify definitions of algorithmic fairness, we will use the following conventions: We will use X to represent a set of individuals, and Y will denote the true set of labels when making decisions regarding each individual in X. The predictively trained machine learning model that we are testing will be Z. In this way, we can verify the main concepts of justice in the topics discussed next.

1.4.1   Fairness Through Unawareness

According to Kusner et al. (2017), an algorithm is fair as long as any protected attributes A are not used explicitly in the decision-making process. Apparently, this definition would be sufficient to guarantee fairness; however, occasionally the nonsensitive attributes in X may contain information correlated to these sensitive (protected) attributes that can lead to discrimination. Furthermore, excluding these attributes can affect the accuracy of the model and generate less effective predictive results.

1.4.2   Individual Fairness

According to Dwork et al. (2012), justice is achieved by the principle that any two individuals who are similar in relation to a specific task should be classified similarly. Thus, an h model with individual fairness, as shown in Eq. (1.1), should provide similar predictive results among similar individuals.

$$P\{x_i \in X\} = P\{x_j \in X\} \Leftrightarrow d(x_i, x_j) < \epsilon \tag{1.1}$$

where d is a distance metric for individuals that measures the similarity between them.

1.4.3   Group Fairness

According to Zhang et al. (2020), if groups selected based on sensitive attributes have the same probability of decision outcomes, this model in theory has justice. There are several types of group fairness; among them we have statistical parity, equalized odds, and equal opportunity. According to Zafar et al. (2017), statistical parity is a popular measure of group fairness. It requires that a decision be independent of the protected attributes. As an example, we can consider that G_1 and G_2 are two groups belonging to x divided by a sensitive attribute $\alpha \in A$, where A is a set of sensitive (protected) attributes. An h model under test has group fairness if it satisfies Eq. (1.2).

$$P\{x_i \in G_1\} - P\{x_j \in G_2\} < \epsilon \tag{1.2}$$

Hardt et al. (2016) proposed another approach to group fairness, related to equalized odds. Here we say that a predictor Y satisfies equalized odds, respecting the protected attribute A and the result Y, if Y and A are independent conditionals on Y. Unlike statistical parity, equalized odds allow Y to depend on A, but only through the variable of target Y. As such, the definition encourages the use of features that allow directly predicting Y, but prohibits abusing A as a proxy for Y. Equalized odds apply to protected targets and attributes taking on values in any space, including binary configurations, multiclass, continuous, or structured.

According to Hardt et al. (2016), when the destination label is set to positive, equalized odds become equal opportunity. This requires that the true positive rate must be the same for all groups. As an example, we can say that a model h satisfies equal opportunity if h does not depend on the protected attributes when a target class Y is set to be positive, as shown in Eq. (1.3).

$$P\{x_i \in G_1, Y = 1\} = P\{h(x_j) = 1 \mid x_j \in G_2, Y = 1\} \tag{1.3}$$

1.4.4 Counterfactual Fairness

According to Kusner et al. (2017), a model satisfies counterfactual fairness if its output remains the same when the protected attribute is inverted to a counterfactual value and other variables modified as determined by the assumed causal model. Let's assume that a is a protected attribute, a' is the counterfactual attribute of a, and x_i' is the new entry with a changed to a'. The model is counterfactually fair if, for any input x_i and protected attribute, a yields Eq. (1.4).

$$P\{h(a \in A, x_i \in X\} = P\{h(x_i')_{a'} = y_i \mid a \in A, x_i \in X\} \tag{1.4}$$

Now that we have talked about different types of fairness definitions, it is important to highlight that it is not possible to use all available definitions of fairness to correct an unfair algorithm. It is necessary to understand the definitions of fairness, analyze the model, and then start a fairness analysis.

1.4.5 Section Summary

This section provided an in-depth exploration of the multifaceted realm of algorithmic fairness. The discussion encompassed various dimensions of defining fairness within algorithmic systems. By addressing these diverse facets of algorithmic fairness, this section contributed to a comprehensive understanding of the intricate considerations involved in designing and evaluating fair algorithms.

1.5 Fairness Metrics

Human prejudice in our society is an extremely sensitive issue and is not always easily defined or identified. On the other hand, prejudice (bias) in the machine learning field is, in general, mathematical. Given this, several different types of tests can be performed on a model to identify different types of bias in its predictions. These tests are defined according to the metric used for algorithmic justice, and their execution depends mainly on the interests and context in which the model is being used. Some examples of fairness metrics are discussed next.

1.5.1 Statistical Parity Difference

According to Dwork et al. (2012), statistical parity or demographic parity is the property that the demographics of those who receive positive (or negative) ratings are identical to the demographics of the population as a whole. Statistical parity is more related to group justice than to individual justice, as it equalizes outcomes between protected and unprotected groups.

Statistical parity suggests that a predictor is unbiased if the prediction y is not dependent on the protected attribute p, as shown in Eq. (1.5).

$$PrPr(p) = Pr(\hat{y}) \tag{1.5}$$

Considering that the same proportion of each population is classified as positive, we can have a result with different false-positive and true-positive rates if the true output of y varies according to the protected attribute p. Deviations from statistical parity can be measured through what we call the statistical parity difference shown in Eq. (1.6).

$$SPD = PrPr(\hat{y} = 1, p = 1) - Pr(\hat{y} = 1, p = 0) \tag{1.6}$$

1.5.2 Equal Opportunity Difference

According to Bellamy et al. (2018), this metric is calculated as the difference in true-positive rates between unprivileged groups and privileged groups. The true-positive rate is the proportion of true positives in relation to the total number of true positives in a given group. A value of 0 means both groups have the same benefit, a value less than 0 means greater benefit to the privileged group, and a value greater than 0 implies greater benefit to the unprivileged group.

Equality of opportunity is satisfied if the prediction y is conditionally independent of the protected attribute p, given the true value of $y = 1$, as shown in Eq. (1.7).

$$EOD = PrPr(\hat{y} = 1, y = 1, p = 1) - Pr(\hat{y} = 1, y = 1, p = 0) \tag{1.7}$$

1.5.3 Average Odds Difference

According to Bellamy et al. (2018), the average odds difference is defined by the average of the difference in false-positive rates and true-positive rates between nonprivileged and privileged groups. A value of 0 implies that both groups have equal benefits, a value less than 0 implies greater benefit to the privileged group, and a value greater than 0 results in a greater benefit to the nonprivileged group, shown in Eq. (1.8).

$$AOD = A_V(TPR, FPR)NPr - A_V(TPR, FPR)Pr \tag{1.8}$$

1.5.4 Disparate Impact

According to Zafar et al. (2017), disparate impact arises when a decision-making system provides results that most often benefit a group of people who share a sensitive attribute value in relation to other groups of people. This metric is defined by the ratio in the probability of favorable outcomes between underprivileged and privileged groups. A value of 1 implies that both groups have equal benefit, a value less than 1 means greater benefit to the privileged group, and a value greater than 1 implies greater benefit to the unprivileged group. We can say that a binary classifier does not suffer disparate impact if satisfies Eq. (1.9).

$$P_R(p = 0) = P_R(\hat{y} = 1 \mid p = 1) \tag{1.9}$$

That is, if the probability that a classifier assigns an individual to the positive class, $\hat{y} = 1$, is the same for both values of the sensitive attribute p, then there is no disparate impact.

1.5.5   Theil Index

Speicher et al. (2018) proposed a metric of justice based on inequality indices in the economy. In their argument, they asserted that general individual justice consists of justice between the group and within the group. The researchers considered a family of inequality indices called generalized entropy indices, including the coefficient of variation and the Theil index, as special cases. The Theil index is the generalized entropy index with $\alpha = 1$, and is calculated according to the following equations to show individual and group fairness. A value of 0 implies perfect justice, shown in Eq. 1.10:

$$b_i = \hat{y}_i - y_i + 1 \tag{1.10}$$

where b_i represents the benefit in the Theil Index, shown in Eq. 1.11.

$$\varepsilon^\alpha(b_1, b_2, \ldots, b_n) = \frac{1}{n} \sum_{i=1}^{n} \left(\frac{b_i}{\mu}\right) Ln\left(\frac{b_i}{\mu}\right) \tag{1.11}$$

1.5.6   Section Summary

This section offered a comprehensive examination of fairness metrics. The discussion encompassed key fairness metrics used to evaluate the fairness of algorithmic systems. By delving into these distinct fairness metrics, this section enhanced readers' understanding of the methodologies for assessing and enhancing algorithmic fairness across different contexts.

1.6   Methods for Fair Machine Learning

According to D'Alessandro et al. (2017), bias mitigation algorithms are intended to improve fairness metrics by modifying the training data, the learning algorithm, or the predictions. In general, these algorithms are categorized into preprocessing, in-processing, and postprocessing. These categories are based on where these algorithms can intervene in a complete machine learning flow. In the case where the algorithm is allowed to modify the training data, preprocessing can be used. If the learning procedure of a learning model can change, then processing can be used. If the algorithm can only treat the learned model as a black box, without any possibility to modify the training data or the learning algorithm, then only the postprocessing can be used.

1.6.1   Preprocessing

Preprocessing techniques try to eliminate implicit discrimination by transforming the data before any modeling. It is possible to use these techniques only in cases where modification of training data is allowed (D'Alessandro et al., 2017). The idea behind preprocessing algorithms is that if the classifier is trained on balanced data, its predictions will consequently be more balanced. According to Bellamy et al. (2018), the preprocessing techniques are as follows: reweighting, optimized preprocessing, learning fair representations, and disparate impact remover.

1.6.2   In-processing

The in-processing method tries to make modifications to the traditional learning algorithms to mitigate discrimination during the model training phase.

If it is allowed to make changes in the training process, the method can be used to incorporate changes in the objective function or impose a constraint (Mehrabi et al., 2021). Adversarial debiasing and bias removal are processing bias mitigation techniques.

1.6.3   Postprocessing

Postprocessing is the last class of bias mitigation methods and can be performed after training. This technique uses a validation set that was not involved in the training process to improve the fairness of forecasts (D'Alessandro et al., 2017). When there is no possibility to make changes to the training data or the model training procedure, only postprocessing techniques can be used. Equalized odds postprocessing, calibrated equalized odds postprocessing, and reject option-based classification are some of the postprocessing algorithms.

1.6.4   Reweighing

Reweighing is a preprocessing solution based on removing the dependency between the protected attribute and the class label. According to the reweighing approach, different weights are assigned to instances so that instances of the unprivileged group with the desirable label get higher weights compared to those labeled as an undesirable class. On the other hand, privileged group objects with the desirable label will have lower weights than the undesirable ones (Kamiran and Calders, 2012). The main objective of reweighing is to adjust the sample weights so that the distribution of the training data matches the distribution of the target population. To use this solution without creating a new bias, it is important to understand the problem and correctly identify the protected attributes and the privileged and unprivileged groups associated with the data.

In the work of Kamiran and Calders (2012), the research is restricted to a binary protected attribute Z and a binary classification problem with attribute classes of destination $Y = [y^+, y^-]$, where y^+ e y^- are representations of desirable and undesirable classes, and objects with $Z = 1$ and $Z = 0$ belong to the privileged and nonprivileged community, respectively. The prediction dependency of a classifier y and a protected attribute Z is defined by Eq. (1.12).

$$p(Z = 1) - p(\hat{Y} = y^+ \mid Z = 0) \tag{1.12}$$

A positive dependency represents that the privileged community object ($Z = 1$) has a greater chance of being labeled as positive than the private group object ($Z = 0$), and vice versa.

1.6.5   Adversarial Debiasing

Zhang et al. (2018) presented a framework for mitigating undesirable biases regarding demographic groups in the training data by including a variable for the interest group and simultaneously learning a predictor and an adversary. A supervised deep learning task is considered in which the mission is to predict an output variable Y given an input variable X while remaining unbiased with respect to some variable Z. We refer to Z as the protected variable. The goal is to maximize the predictor's ability to predict Y while minimizing the adversary's ability to predict Z. This can achieve more accurate predictions that exhibit less evidence of Z stereotyping.

The adversarial debiasing framework proposed by Zhang et al. (2018) depends on opponent training to reduce the latent bias of the trained model. To achieve this goal, several networks are trained in such a way that one can resist predicting the target class and, at the same time, prevent the other from predicting the protected attribute. The satisfaction of justice definitions is theoretically proven (Zhang et al., 2018).

1.6.6   Reject Option-Based Classification

The reject option-based classification is a postprocessing technique that provides favorable results for nonprivileged groups and unfavorable results for privileged groups in a confidence band around the decision threshold. This technique was introduced by Kamiran and Calders (2012).

The reject option-based classification recognizes instances that need to be labeled differently from others using the posterior probabilities produced by probabilistic classifiers. This technique can be used in any probabilistic classifier and does not change the learning algorithm or the preprocessing of the original data.

1.6.7   Section Summary

This section provided a comprehensive exploration of diverse methods employed to achieve fairness in machine learning systems. By covering these diverse methods, this section provided readers with insights into the array of strategies available for achieving fairness in machine learning, addressing biases, and promoting equitable outcomes across different stages of the model's life cycle.

1.7   Tools to Help Detect and Mitigate Bias in Machine Learning Models

Over the years, several industries have integrated systems that use machine learning in the decision-making process for products and services that affect our daily lives. As discussed earlier in this chapter, identifying problems and addressing possible solutions related to algorithmic justice are extremely important. Considering this scenario, some researchers in the AI field have developed tools that help in the identification, reduction, or mitigation of possible biases. In the following sections, we will introduce two different tools that can be used for this purpose.

1.7.1   AI Fairness 360

Bellamy et al. (2018) present in their research an open-source Python toolkit for algorithmic fairness, AI Fairness 360 or AIF360. This toolkit aims to facilitate the transition of fairness research algorithms to use in an industrial setting and to provide a common framework for fairness researchers to share and evaluate algorithms.

The application includes a comprehensive set of metrics for datasets and models to test for biases, explanations for these metrics, and algorithms to mitigate bias. It is possible to learn more about the tool with an interactive web experience at https://aif360.res.ibm.com/.

1.7.2   Aequitas

Saleiro et al. (2019) present in their research an open-source bias and fairness audit toolkit, Aequitas. It is an intuitive and easy-to-use addition to the machine learning workflow, enabling users to seamlessly test models for several bias and fairness metrics in relation to multiple population subgroups.

It is possible to test the tool with preloaded data, or you can upload your own data. You can define the protected groups, select fairness metrics, and generate a bias report. To learn more about the tool you can access http://aequitas.dssg.io/.

1.7.3   Section Summary

This section delved into essential tools designed to identify and alleviate bias in machine learning models. The discussion revolved around two prominent tools dedicated to enhancing fairness. By examining these two integral tools, this section equips readers with insights into practical solutions for detecting and mitigating bias in machine learning models, ultimately contributing to the promotion of fairness and equitable outcomes in various applications.

1.8   Best Practices to Build a Fairer Application

In this chapter, we discussed definitions of fairness, metrics, bias mitigation algorithms, and tools that help to detect and mitigate bias. In addition to these essential topics, it is fundamental to think critically about a possible framework to reduce the algorithmic bias during the process of building a new solution using machine learning. Figure 1.3 provides an example of such a framework.

Figure 1.3 shows a simple generic framework with best practices to build an application that uses machine learning. First of all, when we start designing our system, we need to consider the team that we are putting together to build this model. Diversity of people, stories, and experiences is key to thinking critically about multiple diverse scenarios. This is important because the types of problems that can occur in these datasets and how those problems can cause harm to people in the context of this application are so varied that we don't have a clear sense of how to systematically think about it.

During the design phase, it is important to understand the task and all the objectives of the application. It is also important to identify all stakeholders, the possible errors, and the related consequences. In the data step, it is crucial to verify the dataset. Check the origin of the data and if these data are representative enough in relation to the objective of the final application.

During the model phase, you must check all the machine learning models used and validate all the results. It is important to understand the decisions made by the model and which mechanisms explain the obtained results. Check if the result's evidence is

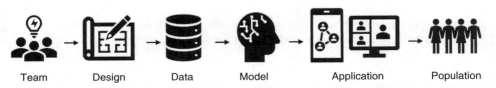

Team   Design   Data   Model   Application   Population

FIGURE 1.3   Best practices to build a fairer application.

Source: Own authorship.

consistent enough. After implementing it in the real world, it is necessary to ensure the quality of the optimization for the application. We need a set of metrics aligned with responsible practices, avoiding damages. These metrics and the performance of the model should be constantly monitored. Build an action plan to identify and respond to failures and damages when they happen. Always try to collect feedback from the population that uses the product built.

1.8.1   Section Summary

This chapter emphasized the importance of critically addressing algorithmic bias during the development of new machine learning solutions. While earlier sections covered fairness definitions, bias metrics, and mitigation algorithms, the focus here is on a comprehensive framework for reducing algorithmic bias. The presented framework, depicted in Figure 1.3, outlined best practices for building machine learning applications.

1.9   Chapter Summary

In this chapter, our aim was to encourage critical thinking regarding the intricate relationship between AI and ethics. We explored the concept of algorithmic justice and examined the five key sources of injustice commonly observed in machine learning. We presented various definitions of algorithmic fairness and introduced fairness metrics that can be employed to identify potential biases within the machine learning process. Furthermore, we discussed a range of bias mitigation algorithms that can be utilized to enhance fairness in machine learning models. Notably, tools such as AI Fairness 360 and Aequitas provide practical solutions for implementing these algorithms. Lastly, we shared a set of best practices for developing applications that leverage AI while promoting fairness. By adhering to these practices, we can strive toward a future where AI technologies are designed with ethics in mind and contribute to a more equitable society.

Chapter Glossary

Term	Definition
Adversarial debiasing	Adversarial debiasing is an in-processing technique that trains a machine learning model to resist and reduce bias during the learning process.
Algorithm	An algorithm refers to a step-by-step set of instructions or rules used to solve a problem or perform a specific task in the context of artificial intelligence and computer science.
Average odds difference	Average odds difference measures the average difference in false-positive rates and true-positive rates between different groups defined by protected attributes.
Bias	Bias in the context of AI refers to the presence of systematic and unfair errors or favoritism in the data or algorithms, leading to unequal treatment of different individuals or groups.

(Continued)

Term	Definition
Disparate impact	Disparate impact refers to a situation where the outcomes of a machine learning model disproportionately affect certain groups based on their protected attributes.
Equal opportunity difference	Equal opportunity difference measures the difference in the true positive rates (sensitivity) between different groups defined by protected attributes.
Fairness	Fairness in artificial intelligence pertains to the equitable treatment of individuals or groups in the development, deployment, and outcomes of machine learning models, free from bias or discrimination.
Favorable label	The favorable label represents the outcome or class considered positive or advantageous in the context of a machine learning task, such as being approved for a loan or being hired.
Group fairness	Group fairness aims to ensure that different groups defined by protected attributes receive equitable treatment and opportunities from machine learning models.
Individual fairness	Individual fairness focuses on ensuring that similar individuals are treated similarly by a machine learning model, irrespective of their protected attributes.
In-processing	In-processing refers to techniques that modify the machine learning model during training to ensure fairness.
Machine learning model	A machine learning model is a computational system that learns patterns and relationships from data to make predictions or decisions without being explicitly programmed for each case.
Metrics	Metrics are quantitative measures used to evaluate the performance and characteristics of machine learning models, including fairness-related measures.
Postprocessing	Postprocessing involves adjusting the model's predictions after training to achieve fairness.
Preprocessing	Preprocessing involves modifying the data before training a machine learning model to mitigate biases or enhance fairness.
Protected attribute	A protected attribute is a characteristic, such as race, gender, or age, that is legally or ethically protected from being used as a basis for discriminatory decision making in AI systems.
Reject option-based classification	Reject option-based classification is a postprocessing technique that incorporates a rejection threshold to reduce the impact of biased predictions on certain groups.
Reweighing	Reweighing is a preprocessing technique that assigns different weights to different data points to mitigate biases.
Statistical parity difference	Statistical parity difference measures the difference in the proportion of favorable outcomes between different groups defined by protected attributes.
Theil Index	The Theil Index is a fairness metric used to measure the level of inequality in predictions or outcomes between different groups.

References

Barocas, S., & Selbst, A. D. (2016). Big data's disparate impact. California Law Review, 104, 671. Retrieved from HeinOnline.

Bellamy, R. K., Dey, K., Hind, M., Hoffman, S. C., Houde, S., Kannan, K., Lohia, P., et al. (2018). AI Fairness 360: An extensible toolkit for detecting, understanding, and mitigating unwanted algorithmic bias. arXiv preprint arXiv:1810.01943.

D'Alessandro, B., O'Neil, C., & Lagatta, T. (2017). Conscientious classification: A data scientist's guide to discrimination-aware classification. Big Data, 5(2), 120–134.

Dwork, C., Hardt, M., Pitassi, T., Reingold, O., & Zemel, R. (2012). Fairness through awareness. In ACM (Ed.), Proceedings of the 3rd Innovations in Theoretical Computer Science Conference (pp. 214–226).

Hardt, M., Price, E., & Srebro, N. (2016). Equality of opportunity in supervised learning. Advances in Neural Information Processing Systems, 29, 3315–3323.

Kamiran, F., & Calders, T. (2012). Data preprocessing techniques for classification without discrimination. Knowledge and Information Systems, 33(1), 1–33.

Kleinberg, J., Mullainathan, S., & Raghavan, M. (2016). Inherent trade-offs in the fair determination of risk scores. arXiv preprint arXiv:1609.05807.

Kusner, M. J., Loftus, J. R., Russell, C., & Silva, R. (2017). Counterfactual fairness. arXiv preprint arXiv:1703.06856.

Mehrabi, N., Morstatter, F., Saxena, N., Lerman, K., & Galstyan, A. (2021). A survey on bias and fairness in machine learning. ACM Computing Surveys (CSUR), 54(6), 1–35.

Saleiro, P., Kuester, B., Hinkson, L., London, J., Stevens, A., Anisfeld, A., Rodolfa, K. T., et al. (2018). Aequitas: A bias and fairness audit toolkit. arXiv preprint arXiv:1811.05577.

Speicher, T., Heidari, H., Grgic-Hlaca, N., Gummadi, K. P., Singla, A., Weller, A., & Zafar, M. B. (2018). A unified approach to quantifying algorithmic unfairness: Measuring individual & group unfairness via inequality indices. In ACM (Ed.), Proceedings of the 24th ACM SIGKDD International Conference on Knowledge Discovery & Data Mining (pp. 2239–2248).

Suresh, H., & Guttag, J. (2021). A framework for understanding sources of harm throughout the machine learning life cycle. In Equity and Access in Algorithms, Mechanisms, and Optimization (pp. 1–9).

Verma, S., & Rubin, J. (2018). Fairness definitions explained. In IEEE (Ed.), 2018 IEEE/ACM International Workshop on Software Fairness (FairWare) (pp. 1–7).

Zafar, M. B., Valera, I., Rogriguez, M. G., & Gummadi, K. P. (2017, April). Fairness constraints: Mechanisms for fair classification. In Artificial intelligence and statistics (pp. 962–970). Proceedings of Machine Learning Research, Volume 54.

Zhang, B. H., Lemoine, B., & Mitchell, M. (2018). Mitigating unwanted biases with adversarial learning. In ACM (Ed.), Proceedings of the 2018 AAAI/ACM Conference on AI, Ethics, and Society (pp. 335–340).

Zhang, J. M., Harman, M., Ma, L., & Liu, Y. (2020). Machine learning testing: Survey, landscapes and horizons. IEEE Transactions on Software Engineering, 48(1).

End of Chapter Questions

1. What are the five main causes of injustice in the area of machine learning?

2. What type of bias can occur during the data generation process?

3. What type of bias may occur after the machine learning model is developed and deployed into the real world?

4. What is the difference between the concepts of individual and group fairness?

5. Define the reject option-based classification technique.

6. What tools can be used to help detect and mitigate bias in machine learning models?

CHAPTER 2

Going Beyond the Technical: Exploring Ethical and Societal Implications of Machine Learning

Brooke Odle, Katherine Finley, Victoria Longfield, and Rodrigo Serrão

Hope College

Question: What does it mean for a machine learning algorithm to be ethical?

Learning Objectives

Upon completion of this chapter, the student should be able to

- Explain the societal and ethical implications of bias in machine learning, given scenarios at the end of the chapter
- Identify at least two methods for evaluating machine learning algorithms, especially those that they develop
- Name the five best practices for enhancing technical communication skills with digital storytelling

Chapter Overview

The purpose of this chapter is to introduce ethical and societal implications of machine learning and their significance and to present practical ways students can work toward mitigating bias in their algorithms and enhance their communication skills to discuss these topics with others via digital storytelling. This chapter reflects changes in computer science and engineering education, with efforts to improve students' ability to work with global markets and design for the needs of all (ABET, 2021). The exercises and application activities at the end of the chapter will also provide opportunities to sharpen one's digital literacy skills. By the end of this chapter, students should be able to identify ethical principles related to computing and coding. Students should also have a greater awareness of how designs and code can impact others, positively and negatively.

2.1 Introduction

In machine learning, it is important to go beyond the technical aspects and consider the ethical and societal implications of the technology developed. Computer scientists, data scientists, and engineers have a responsibility to protect the public and work with high standards via ethical codes.

In *Race After Technology*, Dr. Ruha Benjamin (2019) states:

> whenever we hear the promises of tech being extolled, our antennae should pop up to question what all that hype of "better, faster, fairer" might be hiding and making us ignore. And, when bias and inequity come to light, "lack of intention" to harm is not a viable alibi. One cannot reap the reward when things go right but downplay responsibility when they go wrong.

Technology is not neutral, so one end user may have a different experience than another end user. These experiences may be shaped by the experiences and assumptions of the designers and may also be influenced by the datasets used to train the algorithm, especially if there is bias in the data or the data are not applied to the model appropriately. When there are different outcomes for different users and one or more groups of users is disadvantaged compared to the rest, this is known as *algorithmic bias*.

2.2 Programming Approaches

There are two main ways to program a computer. One way is via direct instructions of tasks to complete, as done with a computer program. This can be facilitated with fundamental concepts like "if" and/or "switch" constructs, loops, functions or routines, arrays, and so on. Another way to program a computer is to give the computer data and let it make rules and predictions based on the raw data sent to it. This latter approach is machine learning, in which the computer "learns" from previous data to make decisions about the future.

Often, programmers include conditions that account for user error. For example, when applying an "if" or "switch" construct, one may include a default option that processes any user-entry errors so that an inappropriate or invalid entry is not inadvertently evaluated by the program. If there are errors in the data or oversights (intentional or unintentional) in the dataset provided, these errors will propagate through the algorithm, impacting the decisions and predictions made. When these algorithms are applied to or deployed to people, these decisions may be discriminatory, resulting in devastating consequences. In the next section, we will learn how algorithmic bias disadvantages some groups over others. To do this, it is imperative to have a foundation for understanding race and socialization.

2.3 Societal and Cultural Implications of Algorithms

According to sociologists, humans are unique species in the animal kingdom because of their capacity to organize life in societies and to express our diverse ways of living through cultures. Think of societies as systems of relationships connecting people via different cultural expressions. This includes languages, values, beliefs, and norms. It also includes food types, clothes, and religions, among many other things (Giddens, 2009). The way individuals transmit culture to others is through socialization. *Socialization* is the process of teaching others (in particular, children and new members of society) everything considered important about a culture. No one is born with a biologically inherited culture. All cultural expressions are learned and taught from one generation to the next via different agents of socialization, such as the family, schools, peers, and media, to name just a few.

2.3.1 Racial Socialization

The way people are socialized also reflects society's history. For example, the United States as a society came into existence due to colonization, land appropriation, and slavery. During U.S. expansion, white settlers, who later became the society's first citizens, came into contact with different ethnoracial groups, including Native Americans, Blacks, Latinx, and Asians. Throughout U.S. history, these racial groups repeatedly received differential treatment for not being white. Collectively, these groups had their lands stolen; were enslaved, segregated, and lynched; treated as second-class citizens; banned from entering the country; put in internment camps; called illegals; and told to go back from where they came from, among other things. Such inferior treatment was legitimized by society's several institutions, such as the government, judicial system, and education system. Even the laws were written in ways that benefited those who identified as whites and disadvantaged everyone else (Lopez, 2006).

Hence, the socialization of white and nonwhite individuals in this environment usually receives racial messages aligned with this historical past, even when the past is selectively used to point to racial progress. Such messages, when accepted and normalized, are reinforced by the dominant culture and different institutions that persistently characterize nonwhite ethnoracial identities as exotic "others." Aside from that, racial socialization can also be used to prevent conversations on these matters because, for some, acknowledging this can result in admitting privilege.

2.3.2 Racism in Its Many Forms

Social and racial inequities are built over time. But they remain in place because of the support of those in the dominant group. For instance, sociologists have different terms to discuss how the process of treating people based on race operates. Racism, for instance, involves not only what someone thinks about other races (*prejudice*) but also the actions toward the reproduction of inequalities (*discrimination*). It can also be manifest through individual interactions or institutional policies (Golash-Baza, 2017). Sociologist Eduardo Bonilla-Silva (2006) argues that after the civil rights movement, racism became much more covert and manifested through narratives of color-blindness, which he called color-blind racism. He found that many whites (this was the racial group he and his team interviewed) used four rhetorical frames to talk about race without appearing racist. The frames were abstract liberalism, naturalization, cultural racism, and minimization of racism. Abstract liberalism, according to Bonilla-Silva, invoked ideas of equal opportunity, individualism, or freedom of choice to explain racial inequality. The naturalization frame was used to explain racial occurrences as if they were natural. Usually, this frame was mentioned to explain segregation as just a matter of individual choices. Cultural racism happened when participants used cultural arguments to talk about minorities' place in society. This frame was used to remove the racist idea of a supposed biological deficiency and replace it with a supposed cultural deficiency. And finally, the minimization of racism frame suggested that "discrimination is no longer a central factor affecting life chances for people of color" (Golash-Boza, 2017). Those complaining about racism should stop because we live in a postracial society. Each of these frames provided those in the study—and in society—a discursive mechanism necessary to appear not racist while ignoring the persistent racial inequities.

All of these forms of racism, biases, and prejudices operate under the umbrella of structural racism, or, as Bonilla-Silva (2006) argues, racialized social systems, which he defines as "societies in which economic, political, social, and ideological levels are partially structured by the placement of actors in racial categories." For our purpose here, it is essential to understand racism not only as an issue related to individuals who overtly mistreat or discriminate against people of color. Racism needs to be understood today from its structural attributes "that originates in the operation of established and respected forces in society, and thus receives far less condemnation than the first type" (Carmichael & Hamilton, 1967). This is the type of racism that creeps into technology.

2.3.3   Why Does This Matter to Students?

Scholars in different academic disciplines (sociology, political science, mathematics, and Internet studies) have recently written about algorithmic racial bias. However, this should have come as no surprise due to the long history of racial oppression mentioned in this section. Nonetheless, just the fact that such biases exist and affect people of color disproportionately should point us to a much bigger problem: the unfamiliarity with how racism works by computer scientists and engineers. Just like people, algorithms are not neutral. Algorithms are "structured by the values of its designer" (Caplan et al., 2018). If the designer or the data have been compromised by societal messages about racial minorities, "the algorithm can inherit that bias" (Ibid).

Engineering and computer science students interested in designing algorithms must go beyond family and media socialization on issues of race and racism to better understand how their implicit bias perpetuates structural racism. Learning this can create awareness and propel the necessary responsibility involved in designing new, more inclusive technologies. Thus, when developing and deploying algorithms in society, engineers, computer scientists, and data scientists have a responsibility to the public. This tenet is inherent in engineering and computing codes of ethics, such as the Institute for Electronics and Electrical Engineering Code of Ethics (IEEE, 2020) and the Association for Computing Machinery Code of Ethics (Gotterbarn et al., 2018). In the next section, we will delve into the critical ethical considerations of machine learning algorithms.

2.4   Ethical Implications of Algorithms

As we learn more about the power of artificial intelligence and specifically of machine learning, we are also increasingly confronted with its sometimes ethically problematic impacts. Many of the most pressing ethical issues resulting from machine learning involve systematic discrimination that unfairly impacts people, often due to features of their identity such as their race, gender, or socioeconomic status. Ethical issues may involve impacts on privacy, autonomy, and access to resources and opportunities, among other things. These biases in machine learning often reflect, codify, reinforce, and sometimes amplify preexisting social biases. These ethical issues arise in roughly three "parts" or "stages" of the machine learning process (Müller, 2020).

Machine learning systems operate on large amounts of data, and many ethical issues may arise in the "first stage" of data collection and preparation. One particularly widespread issue is privacy concerns about the data being collected and, relatedly, who has access to it, what it is used for, who may be disadvantaged by it, and who may profit off of it (Liu et al., 2021). Making ethical judgments about such issues often involves determining issues of *ownership and rights* with regard to the data (Does someone "own" data that is collected from their behavior? Can we ethically buy and sell personal data?) and issues of *consent and coercion* (What counts as informed consent? Is coercion playing a role?). For example, many of these issues arise in cases of apps collecting health-related data from users: Does the user or the app creator own the data measured and collected by the app? Is health-related data the kind of thing that can be ethically bought and sold? Does signing the "terms and conditions" form constitute informed consent if the user is unaware of the potential uses of her data, and is she being coerced by the payment structure of the app (Parker et al., 2019)? Such issues have come up recently for companies and apps such as Facebook, Google, and YouTube. And these issues may be further amplified by aspects of the user such as their socioeconomic or educational background.

Ethical issues may also arise in this "first stage" as a result of the social structures or institutions that the data often captures or relies on. These social structures and institutions may result in biased input data, which may then be further codified and amplified through the machine learning process. Gaining clarity on these ethical issues and how to address them often involves further investigation into the source of the data, paying attention to issues of *rights and justice* (Whose rights may have been violated or undermined by the relevant social structures and institutions? What would constitute just use of this data, if it can be justifiably used at all?), as well as the potential role of such data in the further perpetuation or interrogation of problematic social structures and institutions (Would using this data further codify and reinforce problematic biases? Could it be used to critique and ultimately undermine them?). Many of these issues arise in using data collected by the police or other criminal justice institutions (court systems, prisons) about criminals (e.g., their race, ethnicity, socioeconomic and educational background, mental health and disability status) (Gutierrez & Kirk, 2017). For example, questions surrounding the data used in various "predictive policing" measures and policies include: Were the police departments involved in overpolicing certain communities. And relatedly, are certain individuals less likely to call the police? Will the data be used primarily for punitive or restorative measures, and how will this impact the communities they highlight?

In the "second stage" involving the technical operation of machine learning systems, model learning, ethical issues are often exacerbated by issues of "opacity," meaning the difficulty in understanding the operations of the relevant machine learning system—essentially how it went from input to output. This is relevant both for those people or users impacted and sometimes the developers themselves and can obscure issues of algorithm bias. Ethically analyzing such issues involves reflection on ethical concepts, including those outlined earlier (e.g., privacy, consent, justice) and can be further amplified by ethical issues at the aforementioned "first stage" (e.g., a biased algorithm may develop from biased input data). Examples of algorithms recently found to have been problematically biased include those used by financial institutions to determine access to things like loans. One recent study found that borrowers from minority groups who were "risk-equivalent" were rejected for loans 14 percent more often and charged interest rates nearly 8 percent higher than their counterparts as a result of one such biased algorithm (Bartlett et al., 2022).

In the "third stage" additional ethical issues may arise in "postprocessing," or the application of outputs from a machine learning system. Obviously, ethical issues at the previous two stages can lead to predictable ethical issues at the application stage (e.g., police data and financial algorithms biased against certain groups of people could lead to further overpolicing and harsher sentencing and lack of financial opportunities, respectively). And additionally, even outputs resulting from seemingly unproblematic data gathering and algorithm development can still lead to ethical issues if they are incorrectly calibrated or improperly generalized, as happens in instances of sampling bias. For example, medical research on conditions including autism and heart attacks has often predominantly focused on how those conditions manifest in men; however, it has often been widely assumed to apply equally to men and women. This then has led to increased rates of misdiagnosis and failure to diagnose these conditions in women (Gesi et al., 2021).

Preexisting ethical guidelines for computing, while they may play a more basic and foundational role, are often only a "starting point," considering the level of complexity involved in some of the specific instances outlined earlier. For example, the Association of Computing Machinery Code of Ethics provides the following "General Ethical Principles" (perhaps more accurately understood as "ethical *imperatives*"): (1) contribute to society and human well-being, (2) avoid harm, (3) be honest and trustworthy, (4) be fair and take action not to discriminate . . . (6) respect privacy, and (7) honor confidentiality (Gotterbarn et al., 2018). To further develop an ethical framework for addressing such issues, true ethical principles are needed that, among other things, address many of the concepts and questions noted earlier, as well as how to balance these ethical imperatives against each other and nonethical imperatives (e.g., involving efficiency, the potential for profit). Three candidates for such ethical principles derive from three approaches to ethical reasoning found in consequentialism, deontology, and virtue ethics (Hursthouse, 2013). Very roughly, deontological approaches favor ethical choices that prioritize respecting relevant moral norms, or "rules," as well as individual autonomy; thus, such approaches would tend to prioritize imperatives (6) and (7) as well as favor solutions to the earlier practical examples that minimize coercion and emphasize individual rights and fairness—even at the expense of monetary or other benefits and resources. In contrast, consequentialist approaches favor promoting "the common good" or the greatest good of the greatest number of people above all else, and thus would prioritize imperatives (1) and (2), as well as solutions that maximize benefits to both users and creators—even at the expense of individual rights and confidentiality. And lastly, a virtue ethics approach would prioritize behaving virtuously (and enabling others to do the same) and thus would likely prioritize imperatives (3) and (4) and favor solutions that enabled and supported individuals impacted by machine learning in further developing virtuous habits and capacities.

The existence of such ethical issues in machine learning and their pervasive, real-world impact have also prompted researchers to increasingly focus on ways to predict, eliminate, or mitigate these issues, including developments such as "fair machine learning" and "discrimination-aware data mining" (Hajian & Domingo-Ferrer, 2013). Additionally, more focus is being paid to the development of adequate legal frameworks and protections to address some of these issues, such as the European Commission's Artificial Intelligence Act (2021) and the European Union's General Data Protection Regulation (The European Union, 2018). Increased study and understanding of the ethical questions and concepts at play in issues of machine learning can help further these efforts. The next section will highlight practical tips that can be implemented to mitigate bias while developing machine learning algorithms.

2.5 Approaches to Mitigate Algorithmic Bias

The previous sections of this chapter introduced the ethical and societal impacts of algorithmic bias. Addressing algorithmic bias in machine learning models and computer programs may seem like a daunting task. However, several scholars and computer scientists have proposed methods to mitigate algorithmic bias and promote algorithmic accountability. In this section, we will explore these implemented and recommended practices.

A quick method that students and developers can use to evaluate their algorithms is to conduct the "gut-check," proposed in *Automating Inequality* by Virginia Eubanks (2017). This work focuses on how algorithmic bias and high-tech tools have been used to disenfranchise the poor and working class. To assess the social and economic implications of their work, she recommends that engineers and data scientists conduct a "quick gut-check" by answering the following two questions:

1. Does the tool increase the self-determination and agency of the poor?
2. Would the tool be tolerated if it was targeted at nonpoor people?

If the answer to one or both of these questions is "No," then the algorithm is biased and needs to be modified before it is disseminated to the public.

When evaluating algorithms in industry, a major concern is that machine learning algorithms are highly unregulated. Companies are not held responsible for reporting performance metrics, evaluation results, or any other outcome measures related to their algorithms to a regulatory body, as they are proprietary. Advocates in favor of algorithmic accountability have proposed independent and enforceable technical equality audits, and the Algorithmic Justice League provides a process for requesting audits (Benjamin, 2019). A system also should be in place to ensure rectification when the algorithm harms others (Caplan et al., 2018). In particular, prominent computer scientists and scholars in the field have proposed that there be a federal regulatory body to oversee algorithmic accountability. With respect to access to data obtained from the public to determine Internet search results and develop other types of machine learning algorithms, Safiya Noble envisions a role for public policy in this work. In *Algorithms of Oppression* (2018), she says that "public policy must open up avenues to explore and assess the quality of group and identity information and that is available to the public, a project that will certainly be hotly contested but should still ensue."

While activists are working toward government regulations and more accountability to the public, there has been a cultural change in some companies regarding how machine learning algorithms are documented. In *Glad You Asked*, computer scientist Deborah Raji discussed model cards that she worked on with colleagues during her time at Google (Are We Automating Racism, 2021). A model card is a one-page document that describes how the machine learning model works and lists any questions that reflect ethical concerns. It also addresses the intended use of the model, where the data were obtained, and how the data were labeled in the model. Lastly, the document also provides instructions on how to evaluate the model with respect to performance on different demographic subgroups. Raji recommends that developers focus their evaluations on the groups that are the most vulnerable. In addition to these practical tips, one can also build the awareness of others. The next section introduces digital storytelling as a means of enhancing technical communication skills and integrates all of the topics addressed in this chapter.

2.6 Speak Up: Communicating Ideas with Digital Storytelling

How do computer scientists and engineers communicate with their peers and colleagues? How does society bring important messages and ideas into the public discourse? One of the many ways that ideas are communicated in modern society is through technology and digital media. Digital projects and communication methods such as short-form videos and podcasts are a popular way of sharing ideas, communicating information, and bringing conversation to the forefront of society. Effective communication and digital storytelling are imperative for stories and information to be shared and listened to by a broader audience.

2.6.1 Digital Storytelling

Regardless of the platform selected to share ideas and engage in conversation with and through, digital storytelling best practices will apply and help stories come to life. Five best practices that can be utilized when planning and designing a digital project are the following:

1. Plan out the content in advance for a streamlined story.

   Whether using an outline, storyboard technique, journal method, or any other way of visualizing or writing out the project, a plan is necessary to ensure the story makes sense and is well crafted. Utilizing "signposts" or a "we will learn . . ." technique for the audience on what one is planning on talking about will indicate to the audience that there is a plan, and this helps them navigate what is being communicated.

2. Cite sources.

   The methodology for citing sources will differ greatly depending on the medium of the story. Citing sources in show notes and in the audio is a great way to tell the audience what is informing the project in a podcast. One could also cite sources in the credits for a video project. Citing sources is important to prevent plagiarism, and it also builds credibility.

3. Use music/sound intentionally.

   Any music or sound that goes into a project should be chosen to specifically match the tone and meaning of the digital project. Any sound effects or additional elements should reflect the content manner and how one is communicating the information to the audience.

4. Craft the project to an intended audience.

   Keeping the audience engaged through the project happens through the crafting of the story with them in mind. All of the language, vocabulary terms, and so on should be optimized to the audience's level of understanding and skill level. For instance, if one is going to be discussing machine learning and coding, one would need to make sure the audience understands the code discussed, or one should explain it to them. Never assume that they know something!

5. Sound quality should be clear.

   Finally, regardless of the project type, it is important that any audio included in the project is of decent quality. Clear audio is important to keep the audience engaged and to ensure that they can hear and engage with the content presented.

In addition to these five best practices, it is important to consider how one will engage the audience from the start. National Public Radio (NPR) has wonderful resources for how to begin digital storytelling projects and devices for beginning stories. Regardless of how one decides to start the story, NPR recommends four things for a successful story:

1. Tightly focus the idea.

2. Make that focus clear to listeners.

3. Tell the audience what to expect. (You'll learn XX or discover what happens to this character/place/policy/etc.)

4. Create a sense of movement or momentum.

All these elements will help to guide [the] digital story and help [the] audience engage with the important information that [one is] communicating to them (MacAdam, 2016).

2.6.2 Other Considerations

In addition to the building and creation of the story, there are other considerations to make sure the story can be shared widely. Asking the following questions can ensure that one does not fall into various digital technology traps:

1. Are materials that follow proper copyright law being used?

   Any digital project will need to follow proper copyright guidelines and procedures to use materials that they have not created such as background music or images. Using materials that have either Creative Commons attributions or are in the public domain are the best ways to ensure that one follows proper copyright guidelines. Creativecommons.org is a great place to start searching for Creative Commons licensed materials (Creative Commons, n.d.). Another great idea is to use the advanced search features in Google images and refine the results down to "creative commons licenses" or "commercial." Public domain materials can be found through many different platforms. The most likely source for these types of projects would be stock images or audio files.

2. On which platform will the digital story be shared?

The platform chosen for sharing the story is important from the beginning of the project. Where one decides to post the digital project should inform the technology used to create the project. The platform decision will inform whether one chooses to create a podcast or a video or another media type all together. For example, if a student wants to post on the social media platform TikTok, a short-form video might be the best bet to engage with conversation. This decision would lead one to determine which video platform to use.

No matter how one decides to engage in conversation and communicate ideas and understanding of mitigating bias in algorithms, approaching the project through story and digital storytelling best practices is a great way to begin engaging with the world around oneself. The creation of a podcast or video project, to name a few, are great ways to start conversations around ethics and machine learning. Through these types of projects, one can begin to engage with the world around oneself and perhaps create real change in the way one approaches technology and ethics.

2.7   Chapter Summary

This chapter went "beyond the code" to understand the ethical and societal implications of computing, machine learning algorithms, and technology. Performing test cases to evaluate code without regarding the ethical and societal aspects of deploying the code or technology to the public can have devastating outcomes. Technology is not neutral, and it behooves computer and data scientists as well as engineers to explore the ethical and social aspects of their designs before deploying them to the public. In closing, reflect on the Oath of Non-Harm for an Age of Big Data. Proposed by Virginia Eubanks, this oath is akin to the Hippocratic Oath that doctors pledge.

Chapter Glossary

Term	Definition
Algorithmic bias	Results when the execution of a program provides different users with different outcomes, in which some of those users are disadvantaged compared to the others.
Culture	The entire way of life of a group of people (material and nonmaterial) that acts as a lens through which one views the world and that is passed from one generation to the next.
Socialization	The process of teaching others (in particular, our children and new members of society) everything we consider important about our culture.

Oath of Non-Harm for an Age of Big Data (Eubanks, 2018)
I swear to fulfill, to the best of my ability, the following covenant:

- I will respect all people for their integrity and wisdom, understanding that they are experts in their own lives, and will gladly share with them all the benefits of my knowledge.
- I will use my skills and resources to create bridges for human potential, not barriers. I will create tools that remove obstacles between resources and the people who need them.
- I will not use my technical knowledge to compound the disadvantage created by historic patterns of racism, classism, ableism, sexism, homophobia, xenophobia, transphobia, religious intolerance, and other forms of oppression.
- I will design with history in mind. To ignore a four-century-long pattern of punishing the poor is to be complicit in the "unintended," but terribly predictable consequences that arise when equity and good intentions are assumed as initial conditions.
- I will integrate systems for the needs of people, not data. I will choose system integration as a mechanism to attain human needs, not to facilitate ubiquitous surveillance.
- I will not collect data for data's sake, nor keep it just because I can.
- When informed consent and design convenience come into conflict, informed consent will always prevail.
- I will design no data-based system that overturns an established legal right of the poor.
- I will remember that the technologies I design are not aimed at data points, probabilities, or patterns, but at human beings.

References

Accreditation Board for Engineering and Technology, Inc. (ABET). (2021). Diversity, Equity & Inclusion. https://www.abet.org/about-abet/diversity-equity-and-inclusion/

Algorithmic Justice League. (2023). https://www.ajl.org/

American Auto. (2021, December 13). Spitzer Holding Company, Kapital Entertainment, and Universal Television.

Are We Automating Racism. (2021, March 31). Glad You Asked. https://www.youtube.com/watch?v=Ok5sKLXqynQ&list=PLJ8cMiYb3G5cOFj1VQf8ykNOI0ptuHybc&index=2

Bartlett, R., Morse, A., Stanton, R., & Wallace, N. (2022). Consumer-lending discrimination in the FinTech era. Journal of Financial Economics, 143(1), 30–56.

Benjamin, R. (2019). Race After Technology: Abolitionist Tools For the New Jim Code. Polity.

Bonilla-Silva, E. (2006). Racism without Racists: Color-Blind Racism and the Persistence of Racial Inequality in the United States. Rowman & Littlefield Publishers.

Caplan, R., Hanson, L., Donovan, J., & Matthews, J. (2018). Algorithmic accountability: A primer. Data & Society Research Institute. https://datasociety.net/wp-content/uploads/2018/04/Data_Society_Algorithmic_Accountability_Primer_FINAL-4.pdf

Carmichael, S., & Hamilton, C. V. (1967). Black Power: The Politics of Liberation in America. Vintage Books.

Coded Bias. (2022). 7th Empire Media and Netflix. netflix.com/title/81328723

Creative Commons. (n.d.). When We Share, Everyone Wins. https://creativecommons.org/

Eubanks, V. (2017). Automating Inequality: How High-Tech Tools Profile, Police, and Punish the Poor. St. Martin's Press.

Eubanks, V. (2018, February 21). A Hippocratic Oath for Data Science. Virginia Eubanks (blog). https://virginia-eubanks.com/2018/02/21/a-hippocratic-oath-for-data-science/

Gallimore, A. D. (2021, August 30). Diversity, equity and inclusion should be required in engineering schools' curricula. Inside Higher Ed. https://www.insidehighered.com/views/2021/08/30/diversity-equity-and-inclusion-should-be-required-engineering-schools-curricula

Gesi, C., Migliarese, G., Torriero, S., Capellazzi, M., Omboni, A. C., Cerveri, G., & Mencacci, C. (2021, July). Gender differences in misdiagnosis and delayed diagnosis among adults with autism spectrum disorder with no language or intellectual disability. Brain Sciences, 11(7), 912. https://doi.org/10.3390/brainsci11070912

Giddens, A. (2009). Sociology (6th ed.). Polity Press.

Golash-Boza, T. M. (2017). Race and Racisms: A Critical Approach (2nd ed.). Oxford University Press.

Gotterbarn, D. W., Brinkman, B., Flick, C., Kirkpatrick, M. S., Miller, K., Vazansky, K., & Wolf, M. J. (2018). ACM Code of Ethics and Professional Conduct. Communications of the ACM, 61(7), 106-112.

Gutierrez, C. M., & Kirk, D. S. (2017, July 1). Silence speaks: The relationship between immigration and the underreporting of crime. Crime & Delinquency, 63(8), 926–950. https://doi.org/10.1177/0011128715599993

Hajian, S., & Domingo-Ferrer, J. (2013). Direct and indirect discrimination prevention methods. In B. Custers, T. Calders, B. Schermer, & T. Zarsky (Eds.), Discrimination and Privacy in the Information Society: Data Mining and Profiling in Large Databases (pp. 241–254). Springer. https://doi.org/10.1007/978-3-642-30487-3_13

Hursthouse, R. (2013). Ethics, Humans and Other Animals: An Introduction with Readings. Routledge.

Institute of Electronics and Electrical Engineers. (2020). IEEE Code of Ethics. https://www.ieee.org/about/corporate/governance/p7-8.html

Liu, B., Ding, M., Shaham, S., Rahayu, W., Farokhi, F., & Lin, Z. (2021). When machine learning meets privacy: A survey and outlook. ACM Computing Surveys (CSUR), 54(2), 1–36.

Lopez, I. H. (2006). White by Law: The Legal Construction of Race. NYU Press.

MacAdam, A. (2016, July 26). How Audio Stories Begin. NPR Training + Diverse Sources Database (blog). https://training.npr.org/2016/07/26/how-audio-stories-begin/

Müller, V. C. (2020). Ethics of artificial intelligence and robotics. In Stanford Encyclopedia of Philosophy.

Noble, S. U. (2018). Algorithms of Oppression: How Search Engines Reinforce Racism. New York University Press.

Parker, L., Halter, V., Karliychuk, T., & Grundy, Q. (2019). How private is your mental health app data? An empirical study of mental health app privacy policies and practices. International Journal of Law and Psychiatry, 64, 198–204.

The European Commission. (2021). Artificial Intelligence Act. https://artificialintelligenceact.eu/

The European Union. (2018). General Data Protection Regulation. https://gdpr-info.eu/

The Social Dilemma. (2020). Exposure Labs and Netflix. netflix.com/title/81254224

End of Chapter Activities

The following exercises and activities are intentionally designed as open-ended extensions of the topics covered in this chapter. These exercises and activities provide opportunities to apply the topics addressed in this chapter to engage students in reflective thinking about contemporary algorithm designs and the algorithms they design. They also highlight examples of algorithmic bias documented and presented in television and movies, giving students a chance to reflect on and critique what they observe. By completing these exercises, it is expected that students will have a deeper consideration for social responsibility as they write algorithms that impact people.

Reflective Exercises

1. Learn more about implicit associations by taking a test via Project Implicit (https://app-prod-03.implicit.harvard.edu/implicit/takeatest.html).

   - Listen to this episode of The Sociological Cinema, in which a sociologist and computer scientist discuss human-robotic interactions and the social implications of robotic designs.

   - **Podcast link:** https://www.thesociologicalcinema.com/blog/on-artificial-intelligence-symbolic-interactionism-and-whether-robots-will-take-over-the-world

   - Write a two- to three-sentence summary of the podcast.

   - Who is the audience for this podcast? Can this podcast episode be understood by a lay audience? Why or why not?

   - What types of sources are used to add credibility to the podcast episode? Are these sources referenced during the podcast, in the show notes, or both?

   - Was the story in the podcast structured in a way that kept the audience's attention? If so, what specifically was done to keep the audience's attention? If not, how could the story or report be told differently?

   - Which portion of the conversation was the most interesting? Why?

2. Dr. Evan Peck created Ethical Module Reflections for the CS 1 course at Bucknell University. Even though they were designed for a specific computer science course, students may find these opportunities to apply ethics and reflection to programming helpful activities. As a challenge, apply what this chapter addresses regarding societal implications of computing to these examples, particularly "Developers as Decision Makers."

   Website: https://ethicalcs.github.io/

Algorithmic bias in science fiction

3. Watch the "Oxygen" episode of *Doctor Who*. Identify all fundamental computing concepts noticed with respect to how the robotic space suits track oxygen credits as well as the smart features in the suit that are enabled by artificial intelligence. Review the ACM Code of Ethics. What are some violations of the code observed throughout the episode with respect to how the suits operate? With respect to social and cultural implications of computing, would this smart suit meet Virginia Eubanks' "gut-check"? Why or why not?

Algorithmic bias in situation comedies

4. Watch the "Racial Sensitivity" episode of *Better Off Ted*. What was wrong with the sensors featured in that episode? With respect to race, which employees were negatively affected by the sensors in the building? Based on what was discussed in this chapter, how is the operation of the sensors a form of algorithmic bias? What does this potentially suggest about the individuals who designed the sensors, and what are some ways these problems with the sensors could have been discovered prior to their installation in the building?

5. Watch the pilot episode of *American Auto*. How was the self-driving car tested? What happens when the Black employee walks in front of the car? What reason does the engineer provide for the recognition failure? Based on what was discussed in this chapter, how is this a form of algorithmic bias? Besides humans, what else may not be recognized by the car's sensors and result in an accident? How can the company improve its testing procedure to ensure the car's sensors can better recognize people of different races as well as other objects? In the episode, one of the Black employees claims that the automatic soap dispensers in the restrooms do not work for her. Although another employee tells her that the dispensers are operated via a push button, this example brings to light bias with automatic soap dispensers and sinks. To learn more about these incidences of bias, read the *Scientific American* article titled "Fixing Medical Devices That Are Biased against Race or Gender" by Claudia Wallis (2021) and answer the following questions: What are three ways that bias can be embedded in medical devices? What are some solutions for addressing these biases?

6. The following *Scientific American* article addresses bias with pulse oximeters and how the problem became a major concern when people with darker skin were hospitalized during the COVID-19 pandemic. According to the article, what was causing the problem with the pulse oximeter readings? https://www.scientificamerican.com/article/fixing-medical-devices-that-are-biased-against-race-or-gender/

7. Read the following NPR article titled "When it comes to darker skin, pulse oximeters fall short" by Craig LeMoult (2022), or listen to the audio version of the article, to learn more about how pulse oximeters work and how their design results in bias for people with darker skin. How does a pulse oximeter work? Why do people with darker skin tend to have less accurate readings from pulse oximeters? According to the scientists and engineers highlighted in the article, what are some innovative approaches being used to develop more inclusive pulse oximeter designs? https://www.npr.org/sections/health-shots/2022/07/11/1110370384/when-it-comes-to-darker-skin-pulse-oximeters-fall-short

Algorithmic Bias and Search Algorithms

8. On a piece of paper, write down the first associations (words, images, ideas) that come to mind for each of the following words: "engineer," "scientist," "computer scientist," and "robot." Conduct an online image search for each of these words. What is observed most frequently in the search results for each term? How do the results compare with the original list generated? Students are encouraged to look at the image results for "engineer," "scientist," and

"computer scientist" and note whether or not they see anyone in the results who looks like them. Students should also note how many people in the results look different from them with respect to race, gender, age, and ability status. What do the search results imply about what an engineer, scientist, and computer scientist look like? What do the image results imply about these fields and where people in these fields work? Are any social groups excluded from the results? What are some ways the algorithms may be refined to generate more inclusive results? For advanced reading on search engines and algorithmic bias, read "Algorithms of Oppression" by Safiya Noble.

Digital Storytelling Extensions

9. Make a short video or podcast to communicate individual takeaways from this chapter. Specifically, what are the intersections between algorithms, computing ethics, and society? Who is the audience, and what is the take-home message for the audience?

10. Create a short video or podcast to discuss algorithmic bias depicted in one of the television programs previously mentioned. Who is the audience, and what are the best ways to communicate this information to the audience? What is the main technical issue in the episode? What ethical dilemmas are present with respect to the ACM Code of Ethics? What are the social implications of computing with respect to race, socioeconomic status, or both in the episode? Propose a redesign of the problematic algorithm or technology in the episode. How does the proposed design mitigate the ethical and social concerns previously identified?

Get Involved and Take Action!

If you would like a more active role in increasing awareness of algorithmic justice, consider joining the Algorithmic Justice League (https://www.ajl.org/). The resources available will expose you to additional ethical and societal impacts of algorithms. In addition, you may consider hosting screenings of or panel discussions or fireside chats on documentaries like *Coded Bias* and *The Social Dilemma*.

CHAPTER 3

Social Media and Health Information Dissemination

Fay Cobb Payton, PhD, MBA
North Carolina State University

Xuan Liu, PhD
North Carolina State University

Lynette Yarger, PhD
Pennsylvania State University

Question: Why does undertheorizing magnify ethical issues in health AI/ML among diverse populations?

Learning Objectives

Upon completion of this chapter, the student should be able to

- Identify common uses and applications of social media for delivering sexual, physical, and mental health information
- Compare and contrast the positive and negative impacts of using social media for health information
- Discuss the common challenges in analyzing social media data
- Describe the problem with bias in creating and disseminating health information and implications to machine learning

Chapter Overview

This chapter will model the reach of mental health posts on Twitter, the social networking site rebranded as "X" in July 2023, via the MyHealthImpactNetwork platform. The term "Twitter" will be used to indicate that the analysis of tweets included in this chapter was produced before the changes to the organizational structure, software features, and algorithms initiated after Elon Musk purchased Twitter in 2022. Interactions with the X platform would preclude a longitudinal data capture given the timing used in this chapter. MyHealthImpactNetwork (@MyHealthImpact) is a scholarly health research network focusing on health care and health disparities. The model includes key performance indicators (KPIs), such as reach, to determine the characteristics of posts that generate the most engagement and expand the audience. The analysis includes finding the best time to post based on past performance and identifying key predictors for broadly amplifying health messages. The chapter discusses the ethical considerations when diverse populations are undertheorized in analyzing user communication and their information-seeking interests via Twitter.

3.1 Introduction

With the ubiquitous nature of social media, it is essential that users get accurate and consistent health information from this source. While it is important to understand and research social media and associated KPIs, such as reach and engagement, uncovering patterns to help recognize, filter, and eliminate bias is equally critical. Machine learning techniques such as natural language processing (NLP), sentiment analysis, and topic modeling are often used to analyze the messages, likes, retweets, images, and videos produced by social media users (Ayo et al., 2020; Neethu and Rajasree, 2013). These techniques involve training models on large datasets of Twitter posts to understand and classify language used in them. Common applications of machine learning models include predicting trends or viral topics based on tweet content, facilitating content moderation on social platforms by detecting hate speech, and identifying influential users based on factors like retweets and follower counts. When used responsibly, machine learning techniques offer an efficient and effective means to identify, classify, enhance, and predict health information seeking and sharing behavior on social media. This chapter will discuss social media messaging as a health education campaign and present strategies to mitigate bias and amplify ethical issues concerning underresearched aspects of health data on social media.

3.1.1 Mitigating Bias in Health Information on Social Media

Since its inception in 2006, Twitter has become an impactful channel for communication between mental health organizations and citizens. There is growing evidence that Twitter effectively disseminates mental health information to communities and follows information shared by communities (Hughes et al., 2008). Twitter is also used for detecting and tracking health misinformation (Joseph et al., 2015; McClellan et al., 2017; Naslund et al., 2016) and detecting individuals who may be experiencing a mental health crisis (Jashinsky et al., 2014). However, the structure of how these mental health messages are diffused and their ability to reach audiences is still undertheorized and undermeasured (Bruns & Hallvard, 2014).

As the use of Twitter expands in mental health promotion and communication settings, Neiger and colleagues (2012) argue that metrics and KPIs become increasingly important for assessing the reach and related benefits of posts. On the one hand, a metric, such as the number of posts or tweets, is a single measurement variable in the context of this study. On the other hand, a KPI represents a broader construct, such as insights, exposure, reach, and engagement. Using KPIs helps health organizations to leverage better the time and effort invested in social media messaging (Neiger et al., 2012, 10).

Researching how these data are disseminated and the reach, exposure, impact, and exposure to mitigate bias and then appropriately filter can be used in machine learning by organizations seeking to use social media messaging.

3.2   MyHealthImpactNetwork: For Students by Students

Payton and Kvasny (2016) studied the technology affordances of MyHealthImpactNetwork (Figure 3.1), an online human immunodeficiency virus (HIV) prevention awareness platform designed to reach Black female college students. The platform serves Black women because this group has an expanded risk of HIV transmission. At the same time, Black women use social media regularly to seek information and create content that gives voice to their unique perspectives regarding the prevention of HIV, a stigmatized disease. Perhaps most importantly, the researchers center Black women on college campuses to amplify their experiences, needs, and concerns.

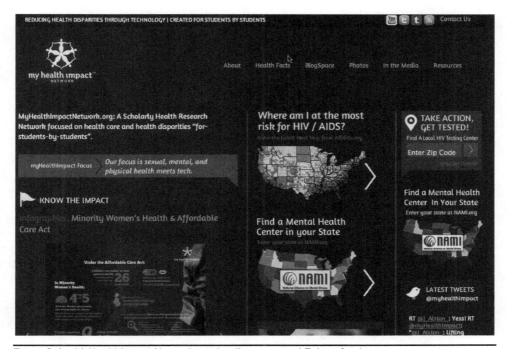

FIGURE 3.1   MyHealthImpactNetwork.org landing page and Twitter feed.

The researchers discovered that Black female college students' use of social media to seek out and share health information is negatively impacted by their awareness of the HIV stigma. The broader societal context that often devalues the lives of Black women further shapes their perceptions of the intended "good" (or affordances) of the MyHealthImpactNetwork platform. Thus, even when intentionally designing software to suit the information needs of minority populations, unintended consequences or uses can yield opposing technology affordances or "goods."

3.2.1   Why Consider Social Media?

Social media has been widely used for mental health communication (Gowen et al., 2012). Berry and colleagues (2017) proposed using the hashtag #WhyWeTweetMH to explore why individuals use the social media website Twitter to discuss mental health. The researchers collected 132 original tweets containing the hashtag #WhyWeTweetMH and identified four themes and eleven associated subthemes depending on the frequency of using common words. The four themes are (1) a sense of community, (2) raising awareness and combating stigma, (3) a safe space for expression, and (4) coping and empowerment. The researchers then discuss the ethical implications of directly applying a Twitter hashtag to generate tweets that are subsequently used for data analysis.

McClellan and colleagues (2017) applied time series analysis and forecasting techniques to identify periods of heightened interest in mental health topics on Twitter. The researchers explored Twitter activities focused on depression or suicide, then forecasted the communication trends over the next 30 days using an autoregressive integrated moving average (ARIMA) model. Their analysis found that spikes in tweet volume following a behavioral health event often last for less than two days, indicating that people circulate mental or behavioral health information associated with heightened periods of interest for a limited time frame.

3.2.2   Studying MyHealthImpact Twitter Feed

This study models Twitter's reach to determine the characteristics of posts that generate the most engagement and expand the audience. A buffer was used to collect data on the @myHealthImpact Twitter account from 08/31/2012 to 12/23/2015. Data collection focused on the presence of terms related to HIV. The data collection resulted in an initial dataset of 1210 observations of 17 variables, where the variables are listed in Table 3.1.

3.2.3   Data Preprocessing

To prepare the dataset for analysis, the "Received Messages," "Social Traffic," "Twitter Posts," and "Web Mentions" variables were removed since they provide little information about the social media or KPIs. Next, the "Received DM" and "Sent DM" features were changed to binary numbers, where "1" means a received direct message (or sent direct message) exists and "0" otherwise. After applying these revisions to the data dictionary, there were 1210 observations of one character variable in the date format, two categorical variables, and ten continuous variables for analyses, as shown in Table 3.1. Since the study aimed to reach more people by posting tweets from @myHealthImpact, there was also a focus on the total number of people who viewed the tweets on the Twitter profile. Therefore, "Twitter Potential Reach" was chosen as the response in the analysis.

Variable Name	Interpretation	Variable Type
Date	Date of data collection	Character in date format mm/dd/yy
Tweets	Number of tweets	Continuous
Clicks	Number of clicks	Continuous
Mentions	Number of mentions	Continuous
New Followers	Number of new followers	Continuous
Retweets	Number of retweets	Continuous
Received Messages	Number of received messages	Continuous
Received DM	Whether received direct messages or not	Categorical (binary)
Sent DM	Whether sent direct messages or not	Categorical (binary)
Unique Users	Number of unique users	Continuous
Followers	Number of followers	Continuous
Following	Number of following	Continuous
Twitter Potential Reach	Number of Twitter potential reach	Continuous
Web Traffic	Number of web traffic	Continuous
Social Traffic	Number of social traffic	Continuous
Twitter Posts	Number of Twitter posts	Continuous
Web Mentions	Number of web mentions	Continuous

TABLE 3.1   Data Dictionary for 17 Variables for Analyses

3.2.4   Check Missing Values

The first step in the data analysis was to check for missing values. After discarding the Date variable, there were 1050 observations of 13 variables in the dataset. We computed the probability of missing values for each feature and chose 15 percent as the threshold to decide if we remove the variable or keep it for imputation. Figure 3.2 provides a visual representation of each variable's missing proportions and the patterns of each variable.

The evaluation of the figure yields the following results:

- "Date," "Received DM," and "Sent DM" are complete.
- "Tweets," "Follower," "Followings," "Twitter Potential Reach (TPR)," and "Web Traffic" were good to impute since their probabilities of missing were less than 15 percent.
- "Clicks" were removed since 30 percent of the observations in the dataset were missing.

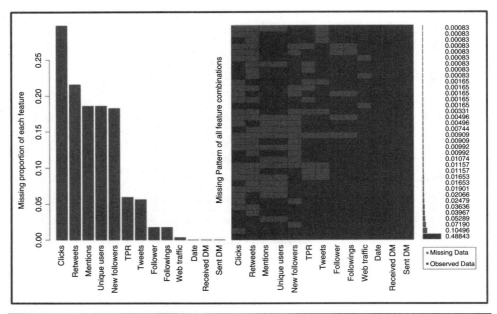

Figure 3.2   Histogram of missing data.

- Since "Mentions," "New Followers," "Retweets," and "Unique Users" were missing in over 15 percent of the observations, they could either be removed or some adjustments applied before imputing them. For example, individuals having at least three missing features could be removed. It was determined that 69 percent (718 out of 1050) of the samples were not missing any information after this adjustment, and the probabilities of missing all features were less than the threshold of 15 percent. As a result, it was safe to apply the imputing procedure.

Next, we imputed these missing values using the Multivariate Imputation by Chained Equations ("MICE") package in R by assuming that the data were missing randomly, and a special margin box plot was used to check this assumption. The number of imputed datasets was set to $m = 50$, which was large enough to reduce the effect of the special initialized seed. Then, the original and imputed data distribution was inspected, as shown in Figure 3.3, where each imputed variable's distributions were checked using a density plot.

The thick curves in Figure 3.3 show the density of the imputed variables for each imputed dataset, while the thin curves show the density of the observed data. Since the shape of the thick curves matches the shape of the thin curves, it can be concluded that the imputed data and observed samples exhibit the same distribution, implying that the imputed values are reliable.

3.2.5   Check Multicollinearity

After imputing missing values, there were 1050 observations of 12 variables in the dataset. The next step is to examine the correlation between the nine continuous variables. The nine continuous variables were tweets, mentions, new followers, retweets, unique

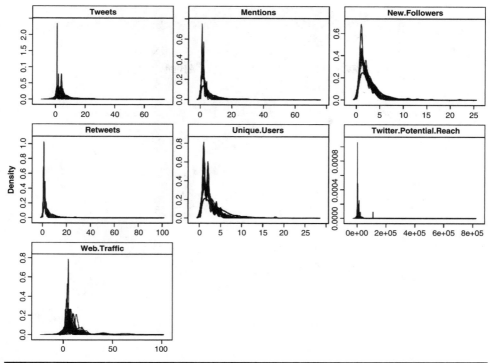

FIGURE 3.3 Imputation diagnosis using density plot.

users, followers, following, Twitter potential reach, and web traffic. Figure 3.4 indicates that "Followers" and "Followings" as well as "Mentions" and "Unique Users" are highly correlated, implying that a Pearson correlation test is needed to test for correlation. If two predictor variables are significantly correlated, one of them should be removed from the analysis to avoid multicollinearity.

3.3 Results of Data Inferential Analysis

3.3.1 Time Series Analysis

A "Twitter Potential Reach" forecast can be conducted using time series analysis. The time series analysis starts by converting the dataset from a data frame to an extensible time series, including "Date" as the time index and "Twitter Potential Reach" as the series. Here, the Buffer application was used to obtain such a time series object. Note that the time index is based on the Coordinated Universal Time (UTC) time zone and has daily periodicity from 2012-08-31 to 2015-12-23 after the transformation.

Figure 3.5 plots the daily total potential reach from 08/31/2012 to 12/23/2015. Next, to provide a clearer view, Figures 3.6 through 3.9 are plots of total potential reach for each year separately.

Since these figures had a large scale, a natural logarithm transformation was applied. Figures 3.10 and 3.11 show a plot of log(reach) and its corresponding autocorrelation Function (ACF) and partial autocorrelation Function (PACF) plots.

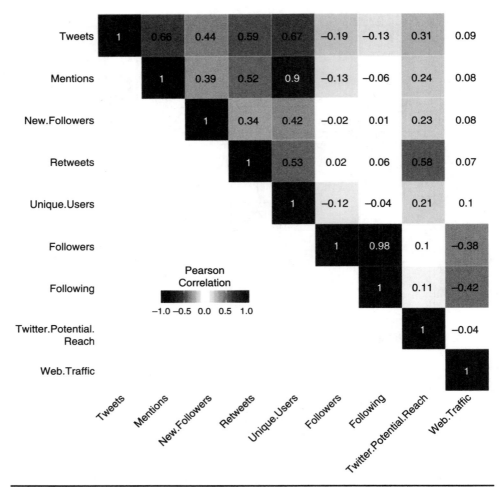

FIGURE 3.4   CorHeatMap of continuous variables.

Since there is an increasing trend in Figure 3.10, the ACF plot in Figure 3.11 decays slowly with data and is still not stationary. A first derivative to log (daily total potential reach) was taken. Seasonality and lags were observable in the data. A cyclical, seasonal pattern occurs and can be affected by social, cultural, and religious events or variations in the length of months. Lags mean a time delay between two time series.

Examining the ACF plot in Figure 3.11 indicates stable positive spikes at lag 7, 14, 21, 28, and so on. The spikes indicate a strong seasonal pattern with seasonal period 7 (weekly period) and suggest taking a seasonal difference with order 7, which is the change from one week to the next. Note that when both seasonal and nonseasonal differences are applied, they yield the same results no matter which is done first. Moreover, since the data have a strong seasonality, the authors decided to take the seasonal differencing first since the resulting series may be stationary, and there will be no need for further nonseasonal differences. Therefore, the authors used a seasonal differencing with period 7 and drew the corresponding plots.

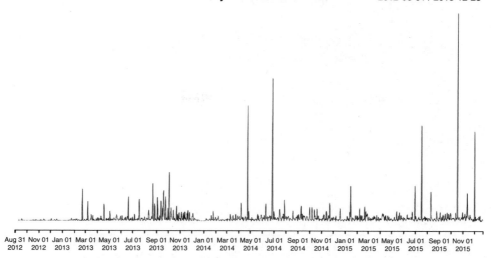

FIGURE 3.5 Plot of daily total potential reach from 08/31/2012 to 12/23/2015.

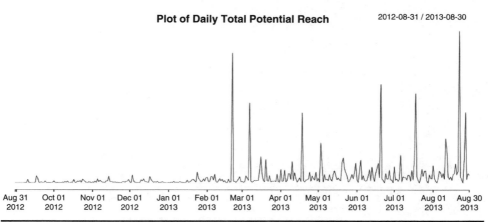

FIGURE 3.6 Plot of daily total potential reach from 08/31/2012 to 08/30/2013.

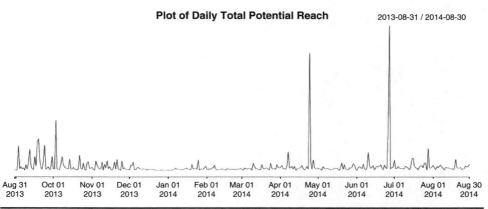

FIGURE 3.7 Plot of daily total potential reach from 08/31/2013 to 08/30/2014.

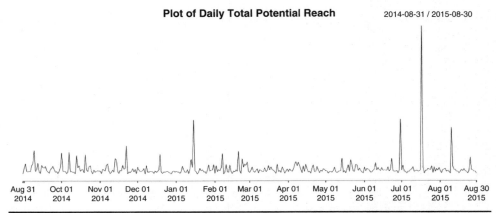

FIGURE 3.8 Plot of daily total potential reach from 08/31/2014 to 08/30/2015.

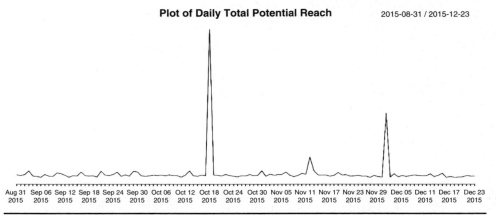

FIGURE 3.9 Plot of daily total potential reach from 08/31/2015 to 12/23/2015.

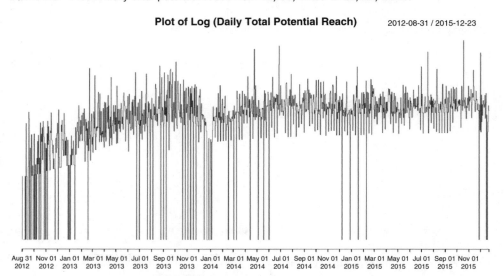

FIGURE 3.10 Plot of log(daily total potential reach).

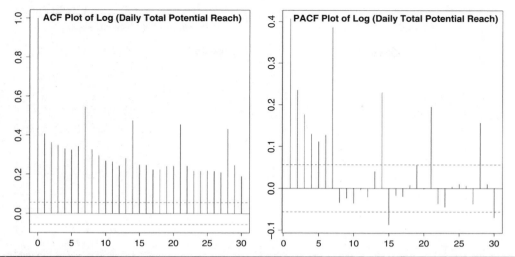

FIGURE 3.11   ACF and PACF plot of log(daily total potential reach).

After the seasonal differencing, we found that the plot of diff(log(reach),7) has no trend in Figure 3.12, the corresponding ACF decays fast, and the PACF tails off in Figure 3.13. Therefore, the series is currently stable. It was possible to check the stationary series by applying the Kwiatkowski-Phillips-Schmidt-Shin (KPSS) test. Since the p-value was greater than 0.05, there was insufficient evidence to reject the null hypothesis, so the current series is stationary and nonseasonal at a significance level $\alpha = 0.05$.

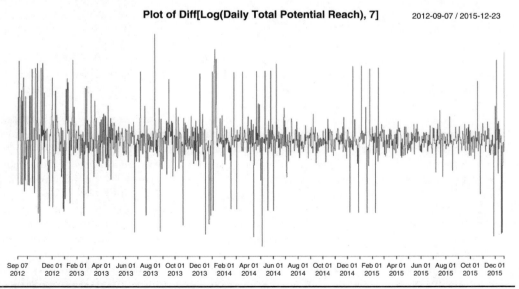

FIGURE 3.12   Plot of Diff [log(daily total potential reach), 7].

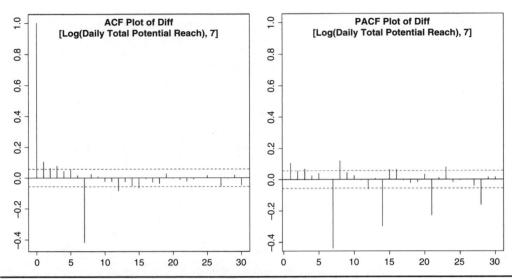

FIGURE 3.13   ACF and PACF plot of Diff [log(daily total potential reach), 7].

3.4   Chapter Summary

Future and additional analyses would permit the determination of the seasonal and nonseasonal components of the model. These components can inform decision makers when social reach and user engagement are more or less likely impacted by social (e.g., Black Lives Matter and protests), public health (e.g., COVID-19), and cultural (e.g., on-campus events based on diversity of cultures) factors. Beyond the quantitative analyses, there needs to be an understanding of technology intended for a social good, which must be paired with context and small data. Namely, Black female and male college students primarily engaged with the @myhealthimpact channel and were the lead delegates in the design of the platform and initial health information dissemination strategy. Data from the Kaiser Family Foundation (Hill et al., 2023) indicates that health information, prevention, and treatment are not equitably provided, designed, or accessed among racial and ethnic groups.

The ethical issues associated with data analyses and interpretation of results are paramount. While the spikes in tweet volume can indicate user interest and engagement with health care providers, nonseasonal periods can best be informed by qualitative methods and users' lived experiences. To this end, @myhealthimpact shifted its focus to be inclusive of mental and physical health as informed by the target population. As described in Yarger and Payton (2018), ethical issues abound, particularly given the perception associated with stigmatized health conditions like mental health and HIV. Privacy, relatability, politics, and campus/organizational climate are considerations when individuals are seeking information and making a platform engagement decision.

The work presented in this chapter lends itself to machine learning to inform and automate content recommendations. To do so, NLP algorithms can be deployed, which enables a computer program to understand human language as it is spoken and written.

Sentiment analysis and content categorization can inform how users "feel" and/or interact with the platform's topics, while the redesign of online material can result. In addition to NLP, linear regression analysis can be used to examine relationships between variables in the model. Support vector machines (SVMs) can help to filter spam or analyze user behavior patterns to detect deceptive activities.

While machine learning can benefit users, algorithmic ethics raises questions about biases in results associated with user engagement and reach. Privacy and security concerns raise the question of trade-offs between health information seeking and personal information exchange. Lastly, a deeper understanding of Twitter users' online versus offline behaviors must be discerned in a manner that does not exacerbate bias and harm. Understanding the time frame among lags and the potential impact of health dissemination is critical. In addition, machine learning techniques can be analyzed to determine social media users' narratives, perceptions of disease conditions, and sentiment analyses as well as informing NLP. This understanding, however, must be informed with comprehension and policies addressing predesign, historical and amplification bias, and power dynamics in care delivery, treatment, and access.

Chapter Glossary

Term	Definition
Data lag	A time delay between two (data) time series.
Data seasonality	Pattern occurs that is cyclical and can be affected by events or variation in the length of months (or a time period).
Health disparities	Preventable differences in the burden of disease, injury, violence, or opportunities to achieve optimal health that are experienced by socially disadvantaged populations.
Health information-seeking	Describes how individuals obtain information about their health, illnesses, and disease conditions.
Natural language processing (NLP)	A subset of machine learning that enables computers to understand, analyze, and generate human language.
Sentiment analysis	Process of using textual data forms to determine emotional tones.
Social media reach	A measure of the number of users/people who see or have been exposed to a post and/or media content.
Support vector machine (SVM)	A linear model for classification and regression problems. It can solve linear and nonlinear problems and is considered supervised machine learning.
Time series analysis	A series of data points or observations recorded at different or regular time intervals, including daily, weekly, or monthly.

References

Ayo, F. E., Folorunso, O., Ibharalu, F. T., & Osinuga, I. A. (2020). Machine learning techniques for hate speech classification of Twitter data: State-of-the-art, future challenges, and research directions. Computer Science Review, 38, 100311.

Berry, N., Lobban, F., Belousov, M., Emsley, R., Nenadic, G., Bucci, S. (2017). #WhyWeTweetMH: Understanding why people use Twitter to discuss mental health problems. Journal of Medical Internet Research, 19(4), e107.

Bruns, A., & Hallvard, M. (2014). Structural layers of communication on Twitter. In K. Weller, A. Bruns, J. Burgess, M. Mahrt, & C. Puschmann (Eds.), Twitter and Society (pp. 15–28). Peter Lang.

Chandler, R., Guillaume, D., Parker, A. G., Mack, A., Hamilton, J., Dorsey, J., & Hernandez, N. D. (2021). The impact of COVID-19 among black women: Evaluating perspectives and sources of information. Ethnicity & Health, 26(1), 80–93. doi: 10.1080/13557858.2020.1841120

Gowen, K., Deschaine, M., Gruttadara, D., Markey, D. (2012). Young adults with mental health conditions and social networking websites: Seeking tools to build community. Psychiatric Rehabilitation Journal, 35(3), 245–250.

Hill, L., Nugga, N., & Artiga, S. (2023, March 15). Key data on health and health care by race and ethnicity. Kaiser Family Foundation. https://www.kff.org/racial-equity-and-health-policy/report/key-facts-on-health-and-health-care-by-race-and-ethnicity/

Hughes, B., Joshi, I., & Wareham, J. (2008). Health 2.0 and Medicine 2.0: Tensions and controversies in the field. Journal of Medical Internet Research, 10, e23.

Jashinsky, J., Burton, S. H., Hanson, C. L., West, J., Giraud-Carrier, C., Barnes, M. D., & Argyle, T. (2014). Tracking Suicide Risk Factors through Twitter in the US. Crisis, 35(1), 51–59.

Joseph, A. J., Tandon, N., Yang, L. H., Duckworth, K., Torous, J., Seidman, L. J., & Keshavan, M. S. (2015). #Schizophrenia: Use and misuse on Twitter. Schizophrenia Research, 165(2–3), 111–115.

McClellan, C., Ali, M. M., Mutter, R., Kroutil, L., & Landwehr, J. (2017). Using social media to monitor mental health discussions – Evidence from Twitter. Journal of the American Medical Informatics Association, 24(3), 496–502.

Naslund, J. A., Aschbrenner, K. A., Marsch, L. A., & Bartels, S. J. (2016). The future of mental health care: Peer-to-peer support and social media. Epidemiology and Psychiatric Sciences, 25(2), 113–122.

Neethu, M. S., & Rajasree, R. (2013). Sentiment analysis in Twitter using machine learning techniques. In 2013 Fourth International Conference on Computing, Communications and Networking Technologies (pp. 1–5). IEEE.

Neiger, B. L., Thackeray, R., Van Wagenen, S. A., Hanson, C. L., West, J. H., & Barnes, M. D. (2012). Use of social media in health promotion: Purposes, key performance indicators, and evaluation metrics. Health Promotion Practice, 13(2), 159–164.

Payton, F. C., & Kvasny, L. (2016). Online HIV awareness and technology affordance benefits for black female collegians – Maybe not: The case of stigma. Journal of the American Medical Informatics Association, 23(6), 1121–1126. doi: http://dx.doi.org/10.1093/jamia/ocw017

Yarger, L., & Payton, F. C. (2018). Managing hypervisibility in the HIV prevention information seeking practices of black female college students. Journal of the Association for Information Science and Technology, 69(6), 798–806.

End of Chapter Questions

1. Identify three common uses and applications of social media platforms or apps to deliver mental health information and support to college students.

2. List common arguments about social media's positive and negative impacts on youths and young adults.

3. Given these debates, explain your position on the usefulness of using social media as a platform for mental health promotion and communication.

4. Discuss three common challenges in analyzing social media data presented in the chapter.

5. Explain the "bias problem" in creating and disseminating mental health information.

6. Argue for or against the importance of tailoring mental health information for vulnerable populations.

CHAPTER 4

Comparative Case Study of Fairness Toolkits

**Keith McNamara, Jr., Kiana Alikhademi,
Brianna Richardson, Emma Drobina,
and Juan E. Gilbert**

University of Florida

Question: How effective are fairness toolkits at detecting and mitigating bias and ensuring fairness in machine learning? And what are their shortcomings?

Learning Objectives

Upon completion of this chapter, the student should be able to

- Identify the most common types of biases in the machine learning life cycle through examples

- Explain the objective of fairness in machine learning and understand the nuance of the definition space

- Evaluate state-of-the-art fairness toolkits based on different criteria to find the capabilities and shortcomings

Chapter Overview

In this chapter, we discuss methods to detect and mitigate bias through the use of fairness toolkits, which can be used when building machine learning (ML) models, to ensure they serve the intended populations effectively. Section 4.3 introduces the rise of ML in various domain spaces and highlights the importance of recognizing and mitigating bias using recorded statistics. Section 4.4 details three types of biases you might encounter when working with different datasets along with scenarios of how these biases can be negatively impactful. Section 4.5 highlights how current practitioners implement fairness through the use of a sensitive attribute and how prior work has categorized three different measures of fairness. Section 4.6 introduces the field of responsible artificial intelligence (AI) as a response to various instances of models or datasets found to be biased and failing to include fairness measures. This section also details some existing strategies that practitioners have undertaken to address the issue of fairness. Section 4.7 presents the results of the comparison evaluation of the three fairness software toolkits to outline their capabilities, and Section 4.8 discusses how these results demonstrate not only what they are capable of but, more importantly, also what needs further improvement.

4.1   Introduction

ML is emerging across new industries at a rapid rate, used to assist in business processes, automate mundane tasks, and optimize products with its predictive power. However, as the number of ML applications has risen, the number of AI-related controversies has risen as well. Examples include Sweeney's 2013 findings that Google is more likely to advertise for criminal background checks or mugshot viewing sites on searches for Black-sounding names than white-sounding names, regardless of true criminal background; Buolamwini and Gebru's 2018 discovery that facial recognition systems had error rates of 35 percent when used on dark-skinned women, as opposed to <1% for white men; and ProPublica's 2016 analysis of software for predicting criminal recidivism that misclassified Black defendants as high risk at twice the rate of white defendants.

> **CONSIDER THIS:**
> If researchers and developers are using biased data to build algorithms that are used all around the world, the impact of that bias will affect people across many countries. Furthermore, the impact of bias can be exacerbated in underrepresented communities. We must apply responsible AI to prevent the unintended consequences on different populations.

4.2   Bias

The detection of bias and the implementation of fairness strategies are two distinct steps in the model development process. Selecting the best fairness strategy is highly context-dependent. To better propose and implement fairness strategies, it is critical to understand the source of bias. There are several examples including label bias, sampling bias, and representation bias.

4.2.1   Label Bias

Label bias occurs when the labels we are using are inaccurate or missing information. For supervised ML, we assume the labels are correct and are our "ground truth" that all other decisions will be based upon. Figure 4.1 shows two examples of mislabeling datasets in ML and how it could impact the final outcome.

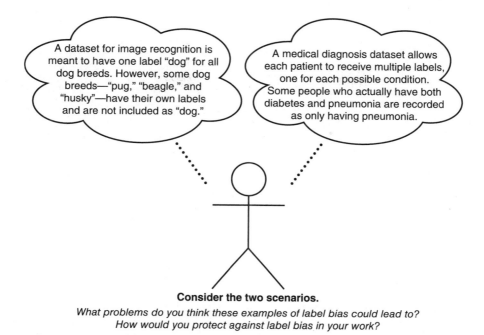

Consider the two scenarios.

What problems do you think these examples of label bias could lead to?
How would you protect against label bias in your work?

FIGURE 4.1   Two different scenarios regarding labeling bias.

4.2.2   Sampling Bias

Sampling bias occurs when certain groups within a population are more likely to be included in a sample than others. If we base our conclusions off of a sample that isn't truly representative of the population, our results will be skewed. Figure 4.2 shows an example of sampling bias where different groups do not have the same likelihood to be selected into the study population.

4.2.3   Representation Bias

Representation bias occurs when the sample used for our task is not representative of the individuals for whom the solution was built. Figure 4.3 illustrates an example of scenarios where the population is not representative of the target audience.

4.3   Fairness

Once we identify the source of bias, we need to mitigate the bias utilizing a fairness measure, which is dependent on use case and objective. Many fairness measures in ML rely on the concept of a **sensitive attribute**, the trait we want to make sure is not influencing our decision making. Often, this is a legally protected class, such as race, age, religion, or sex. Verma and Rubin (2018) divided fairness measures into the following three categories: statistical measures, similarity measures, and causal reasoning.

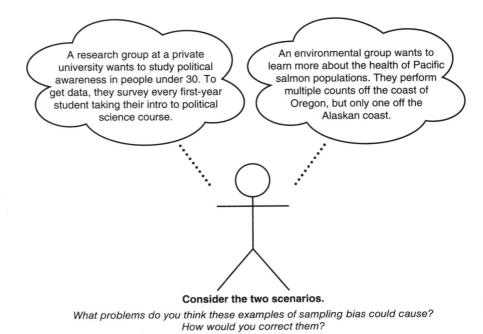

FIGURE 4.2 Two different scenarios of sampling bias.

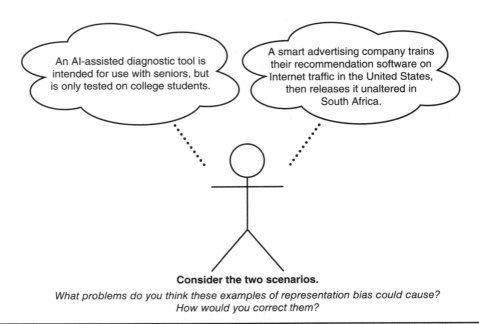

FIGURE 4.3 Two different scenarios of representation bias.

4.3.1 Statistical Measures

Statistical measures measure fairness by comparing the outcome predicted by the algorithm to the actual outcome across a sensitive attribute.

4.3.2 Similarity Measures

Similarity measures compare nonsensitive attributes, under the principle that a fair decision-making system should give the same result for two people with similar nonsensitive attributes but different sensitive attributes.

Causal reasoning determines if a sensitive attribute has an influence on outcome. Typically, this is done by modeling the relationships between attributes as a set of equations.

4.4 Applying Responsible AI

Controversial AI has led to a new branch of research, responsible AI, which focuses on the ethical investigation and application of ML in practice. Responsible AI objectives focus on increasing the transparency, interpretability, explainability, and fairness of algorithms (Cheng et al., 2021). This chapter focuses on the fairness tenet of responsible AI.

It is commonplace to find ML-implementing institutions putting in place their own guidelines and mission statements to address unethical practices in AI via responsible AI objectives (Jobin et al., 2019). While the solution techniques might differ, the main goal for responsible AI remains the same: to produce algorithms that are free from bias (Richardson and Gilbert, 2021). Therefore, fair AI research consists of a diversity of solutions that can be used to detect and/or mitigate bias (Alikhademi et al., 2022). A multitude of different strategies have arisen in this space, and to assist practitioners in implementing fair AI, tools have been produced to condense this research into usable interfaces. These efforts usually arise in the form of checklists or software toolkits.

4.4.1 Checklists

Checklists are guides made for practitioners to ensure the inclusion of ethical practices throughout the pipeline. An example of a checklist is the one made by researchers at IBM to evaluate if ethics for AI is properly embedded in the system design or not (IBM, 2022).

4.4.2 Software Toolkits

Software toolkits are packages that can be imported into the ML life cycle and, similarly, can assist in evaluating bias and producing fair algorithms (Richardson & Gilbert, 2021). Nonetheless, these tools are rarely used in practice. Previous research suggests a disconnect between tool creation and practitioner needs (Holstein et al., 2019; Richardson et al., 2021; Veale et al., 2018). This work aims to investigate the effectiveness of three popular fairness toolkits for supervised tasks and thoroughly discuss the strengths and weaknesses of each.

> **CONSIDER THIS:**
> Developers of these algorithms can also engage in an unconscious bias. This bias is often unintentional but produces the same effect of not considering a diversity of data. Picture someone who is programming an algorithm for a camera to detect a person. If they only consider people of one race, people without disabilities, and so on, then the system will not always recognize alternative data. The result will be people who do not match the information the model was built on being left out when they attempt to use the system.

4.4.3   Fairness Software Toolkits

Fairness software toolkits, as stated earlier in the chapter, are packages that can be imported into the ML pipeline and used to evaluate fairness programmatically. While fairness software toolkits assist ML practitioners in identifying and assessing their algorithms according to standard fairness metrics, the literature suggests that much work is needed to optimize these toolkits for their intended audience. For one, not every toolkit can detect every kind of bias. Most importantly, however, they require thoughtful human operators to determine sensitive attributes, evaluate the data, and investigate the results of different notions of fairness. Still, developers are in the driver's seat to find biases in complex data which could impact the whole process. In the following section, we will teach you about three prominent fairness software toolkits and evaluate them using a rubric of design needs proposed in the literature (Richardson et al., 2021).

Much research has been done to evaluate the efficacy of the fairness software toolkits using interviews and focus groups (Holstein et al., 2019; Richardson et al., 2021; Veale et al., 2018). The results show that practitioners found misalignment between the technical community of practitioners and fairness experts in industry or academia. Practitioner responses provide a critical foundation for outlining design needs for fairness experts to utilize in the creation of fairness toolkits. Major themes in responses suggest practitioners:

- Prioritize explanations and transparency of their models
- Need effective tools to communicate model utility and performance to nonexperts
- Are concerned with the gamification of responsible AI pursuits (Veale et al., 2018).

Furthermore, while fairness toolkits did have significant impact on subsequent decisions, practitioners struggled with gaining valuable insight from fairness results (Richardson et al., 2021). Richardson and Gilbert (2021) identify the institutional gaps that prevent the effective implementation of fairness in organizations, and they provide suggestions for how practitioners can best implement fairness in their own projects.

For the sake of this evaluation, we chose a representative subset of fairness toolkits. The toolkits used in this study include Aequitas (Saleiro et al., 2018), Fairlearn (Bird et al., 2020), and IBM Fairness Toolkit (AIF360) (Bellamy et al., 2018). This subset of toolkits was selected for the diverse set of features offered by each and their easy installation into our Python pipelines.

4.4.4   Evaluating Fairness Toolkits

Using the rubric first proposed by Richardson et al. (2021), we walk through an approach of comparing these three fairness toolkits to evaluate their strengths and weaknesses in addressing bias and fairness. In the original work, authors separate criteria into two major categories: criteria for supporting fairness analysis and criteria for optimal tooling (Richards et al., 2021). The former category summarizes the baseline expectations of fairness toolkits based on fairness literature, focusing on ensuring toolkits have a diverse array of interventions that have been previously proposed in the literature. The latter category focuses on optimizing toolkits for usability and providing users with support in addition to the standard fairness expectations (Richardson et al., 2021).

The following list explains how it was determined if a toolkit satisfied a criterion. In Table 4.1, we provide the rubric criteria and how each toolkit satisfies those criteria to compare their ability to prevent bias and improve fairness in datasets.

- *Applicable to a diverse range of predictive tasks*: Toolkit accepts the outcome of at least two types of tasks from binary classification, multiclass classification, regression, clustering, and ranking.

- *Applicable to a diverse range of data types*: Toolkit works with at least two types of data including tabular, image, and sequence data (text or time series).

This table shows which criterion is supported by which toolkits. Overall, IBM AIF360 and UChicago's Aequitas supported the most and least criteria, respectively.

Criteria	UChicago's Aequitas	IBM's Fairness 360	Microsoft's Fairlearn
Applicable to a diverse range of predictive tasks		✓	✓
Applicable to a diverse range of data types		✓	✓
Applicable to a diverse range of biases			
Inclusive of diverse measures of fairness		✓	
Can detect bias	✓		
Can mitigate bias		✓	
Can intervene at different stages of ML life cycle		✓	✓
Model agnostic	✓	✓	
Fairness criteria agnostic		✓	✓
Performance criteria agnostic		✓	✓
Provides intersectional analysis	✓		
Applicable to data without sensitive features			

TABLE 4.1   Comparison of Aequitas, Fairlearn, and AIF360 Based on the Previous Rubrics Inspired from (Richardson et al., 2021)

- *Applicable to a diverse range of biases' criteria*: We evaluate this with respect to preexisting biases such as representation bias, technical bias such as data processing bias, and emerging bias such as belief bias, which is introduced by Richardson and Gilbert (2021). A toolkit satisfies these criteria if it detects and/or mitigates a bias from each of these categories.

- *Inclusive of a diverse range of fairness*: We consider three categories that we discussed in the background section, including statistical measures, similarity measures, and causal reasoning–based measures. A toolkit satisfies this criterion if it proposes at least one solution from all of these categories.

- *Can detect bias*: The toolkit can detect a bias from each of the preexisting categories outlined earlier.

- *Can mitigate bias*: The toolkit contains measures to mitigate bias from each of the preexisting categories outlined earlier.

- *Can intervene at different stages of ML life cycle*: The toolkit allows for the fairness techniques to be used during preprocessing, in-processing, or postprocessing.

- *Model agnostic*: The toolkit is not constrained to the use of a specific model type (classification, regression, etc.), but can apply the fairness techniques to any model being used.

- *Fairness criteria agnostic*: The toolkit contains functionality to allow the user to enter the fairness metric of their choice. These agnostic metrics can be used for subsequent analysis.

- *Performance criteria agnostic*: The toolkit contains functionality to allow the user to enter the performance metric of their choice. These agnostic metrics can be used for subsequent analysis.

- *Provides intersectional analysis*: The toolkit contains built-in functionality to take in at least two sensitive attribute columns and perform subsequent fairness analysis on intersectional identities.

- *Applicable to data without sensitive attributes*: The toolkit can perform some type of fairness analysis without being given sensitive attribute information.

4.5   Results

Let's look at the results found using each of the toolkits. Table 4.1 represents a rubric created based on available features and capabilities of each toolkit. In the following subsections we will discuss more about each criterion and how the specific toolkit stands for it.

UChicago's Aequitas (2018) is a toolkit for auditing LM models based on a standard set of fairness definitions. It was developed by researchers at the University of Chicago to simplify the auditing process for non-ML researchers. Aequitas could be used for bias detection, fairness assessments, and visualizations. It can be accessed via a web interface or a Python package (Saleiro et al., 2018).

There are many benefits and advantages to using Aequitas. It supports a diverse range of ML tasks, including supervised tasks (such as classification and regression) and unsupervised tasks (such as clustering and time series forecasting). Furthermore, as a bias and fairness auditing toolkit, Aequitas can be used to find a diverse array of biases in the input data and the predictions. It further assists the practitioner by providing diverse plotting and evaluation methods to analyze different bias and fairness metrics across single or multiple attributes. Furthermore, Aequitas is model agnostic, as it provides analysis given outcomes of any type of model, which is beneficial considering the diverse array of ML models that exists. Aequitas also provides a quintessential example of intersectional analysis. It provides the ability to define multiple sensitive attributes and provides auditing results for each subgroup. It also provides the user the opportunity to define a reference group to compare all other subgroups to. Aequitas's major strengths reside in its user-intuitive interfaces and its intersectional analysis.

There is still much to be desired when it comes to Aequitas. Aequitas is only applicable to tabular data, it requires sensitive attributes to conduct analysis, and it does not have a user-friendly method to generate custom fairness and performance criteria. While it provides statistical measures for fairness, it still lacks similarity-based or causal reasoning measures. Furthermore, it works as an auditing toolkit and provides no methods for mitigating the detected bias. Lastly, Aequitas is a post-hoc toolkit and is only used once a model has been trained.

Fairlearn (Bird et al., 2020) is an open-source fairness toolkit created by a team of researchers at Microsoft. Fairlearn focuses on detecting and mitigating fairness concerns in regression and classification models. Fairlearn prioritizes group fairness, aiming to identify disparities that exist between outcomes for sensitive groups. It also includes an interactive interface where users can compare models. Fairlearn can be accessed via a Python package that includes assessment and detection functions, example datasets, and a number of tutorials (Bird et al., 2020).

Fairlearn is a great toolkit with several unique features and advantages. Unlike Aequitas, Fairlearn provides a number of mitigation methods to assist practitioners with satisfying statistical measures of fairness that can be used during preprocessing, in-processing, or postprocessing. Preprocessing methods aim to remove existing biases or correlations across samples. In-processing methods provide fairness constraints to assist during the creation of the model. Lastly, postprocessing methods can be used to process biased outcomes from the model. Fairlearn is model agnostic as long as the given model has a fit() and predict() functionality for building and training the model. However, these functionalities are only needed for in-processing methods. Fairlearn also has an interactive interface to assist in side-by-side comparisons of models and their fairness and performance outcomes. Fairlearn's major strengths reside in this interface and its diverse selection of mitigation methods.

Despite these advantages, Fairlearn still has many avenues for growth. While the creators note the possibility of future updates, currently it is only applicable to binary classification and regression tasks. Furthermore, it is limited to tabular or time series data that contain sensitive attributes. Fairlearn does not provide many options for bias detection, only mitigation. Unlike Aequitas, there is no built-in functionality to handle multiple attributes for intersectional analysis. Lastly, there are no simple methods to incorporate custom fairness or performance measures into the mitigation methods, drastically reducing the control the practitioners have when utilizing these mitigation methods.

IBM's AI Fairness 360. (AIF360) (Bellamy et al., 2018) is an open-source fairness toolkit produced by a team of researchers at IBM. Similar to Fairlearn (Bird et al., 2020)

AIF360 works to detect and mitigate biases. Uniquely, it proposes solutions across both group and individual fairness standards. AIF360 provides a surplus of resources where users can become familiar with its capabilities and shows examples of how users can incorporate popular toolkits from explainability, such as Local Interpretable Model-agnostic Explanations (LIME), (Ribeiro et al., 2018) into their fairness analysis. AIF360 can be accessed via Python or a R package (Bellamy et al., 2018).

AIF360 is one toolkit in a host of toolkits provided by IBM. While it has several strengths as a stand-alone system, it can be used in concert with other toolkits for explainability, privacy, or robustness, amplifying its usefulness. First, out of the example toolkits, AIF360 provides the largest number of metrics for bias detection. Another unique aspect of AIF360 is that it provides fairness measures for both individual and group fairness. Furthermore, it allows for custom fairness and performance metrics. Also unique is that AIF360 provides analysis options for a diverse selection of data types. Similar to Fairlearn, AIF360 provides a number of mitigation methods that can be used for preprocessing, in-processing, or postprocessing. The major strengths of AIF360 include the extensive documentation and resources to assist practitioners in utilizing their toolkit.

AIF360 also has many limitations. Like Fairlearn, AIF360 is limited to supervised tasks, provides no inherent support for intersectional analysis, and is model agnostic to supervised models that contain specific function names.

While the example toolkits each have their own strengths and weaknesses, in totality, they depict general trends of the landscape of fair AI toolkits. Several global limitations emerge when these toolkits are assessed side-by-side.

4.6   What Are the Limitations of These Toolkits?

Next, we provide an overview of what the toolkits accomplish regarding bias, and fairness techniques will be highlighted. Using the criteria of the rubric presented earlier, we provide an explanation of the limitation that each of the toolkits faces.

4.6.1   Diverse Range of Biases

As the sources of these biases differ, the approaches to mitigate and resolve them are different, too. Most of the toolkits discussed in this chapter covered the biases from the preexisting category (Richardson et al., 2021). However, none of the toolkits consider the technical biases or emerging biases. There is a definite need to incorporate more diverse bias detection and mitigation strategies in these toolkits.

4.6.2   Diverse Measures of Fairness

Prior research has shown that one fairness definition might not work in all different scenarios (Alikhademi et al., 2022; Noble, 2018). The majority of the toolkits analyzed in this chapter covered statistical measures including demographic parity, disparate impact, equality of odds, and error rate. Fairness metrics based on similarity measures and causal reasoning could provide more information about the underlying cause of biases, which could play a huge role in detecting and resolving the biases. However, these two types of fairness metrics are completely missed in the toolkits studied here.

4.6.3   Bias Detection

All of these toolkits compute the fairness metrics to help researchers in bias detection. However, none of them detect clearly which type of bias is missed. Aequitas provides a bias report that shows the disparities or metrics failed according to the data. However, this report could only be generated using the web portal and not with the application programming interface (API). If more explicit definitions of biases are incorporated into these toolkits, we can detect biases more clearly and provide detailed reports about the biases.

4.6.4   Bias Mitigation

Any fairness toolkit should analyze the data and model to find biases and perform some operations to remove them. Without proper bias mitigation, these toolkits remain a checklist where the burden of resolving biases is still on the researchers. Many popular bias mitigation techniques could be incorporated into these toolkits.

4.6.5   Intersectional Analysis

Existing biases could be attributed to multiple protected attributes at the same time. An intersectional analysis of protected attributes against the fairness metrics would help find the actual unprivileged groups, and it could help more in alleviating the biases. Among the toolkits we analyzed, only Aequitas lets the researcher analyze the protected attributes for specific metrics simultaneously.

4.6.6   Applicable to Data Without Sensitive Information

Race, sex, age, and disability status are protected attributes defined by law (The National Archives, 2010). However, there could be cases in which data do not contain this sensitive information explicitly but are still biased. For instance, a neighborhood or ZIP code identifier is often correlated to attributes such as socioeconomic status or race, leading to inherent biases. Therefore, each fairness toolkit should look at the data and detect biases even if they do not include the traditional sensitive information such as age, sex, and race.

4.7   Chapter Summary

To further explore the societal impacts of AI , this chapter analyzed some of the most popular fairness toolkits and highlighted their limitations. These fairness toolkits need to incorporate a more diverse range of bias definitions and fairness metrics to become applicable in real-world scenarios. Moreover, none of the tested toolkits allowed for easily implementable intersectional analysis across sensitive attributes, which is a necessary functionality to help researchers find the root cause of the biases in the data and model. The majority of these toolkits need to implement a better method for analyzing the sensitive attributes holistically. Lastly, these toolkits need to incorporate high-level research in the bias mitigation techniques to address practitioners' needs better.

Acknowledgments

Research reported in this chapter was supported by the University of Florida Informatics Institute Fellowship Program.

Chapter Glossary

Term	Definition
Bias	A tendency to favor one person over another one. Bias in machine learning refers to a situation where an algorithm produces results in favor of a specific individual or group.
Causal reasoning	Determines if a sensitive attribute has an influence on outcome.
Checklists	Guides made for practitioners to ensure the inclusion of ethical practices throughout the pipeline.
Emerging bias	Biases that arise from the way the machine learning model is deployed.
Fairness	The quality or state of being fair. Fairness in machine learning refers to efforts in combating bias in machine learning outcomes or models.
In-processing	Procedures occur during the training phase of a machine learning solution.
Postprocessing	Procedures are performed on the final results of machine learning.
Preprocessing	Procedures are applied to raw data before passing them to machine learning systems.
Representation bias	Occurs when the sample used for our task is not representative of the individuals for whom the solution was built.
Responsible AI	A subfield of machine learning focused on building ethical and conscientious machine learning algorithms by upholding a number of tenets, including transparency, interpretability, explainability, and fairness.
Sampling bias	When certain groups within a population are more likely to be included in a sample than others.
Similarity measures	Compare nonsensitive attributes, under the principle that a fair decision-making system should give the same result for two people with similar nonsensitive attributes but different sensitive attributes.
Software toolkits	Packages that can be imported into the machine learning life cycle and can assist in evaluating bias and producing fair algorithms. These tools are rarely used in practice.
Statistical measures	Measure fairness by comparing the outcome predicted by the algorithm to the actual outcome across a sensitive attribute.
Supervised learning	Uses labeled data to predict or classify outcomes.
Technical bias	Biases that arise from technical constraints or issues related to how algorithms or models are designed.
Unsupervised learning	A method for analyzing and discovering patterns in unlabeled data.

References

Alikhademi, K., Drobina, E., Prioleau, D., Richardson, B., Purves, D., & Gilbert, J. E. (2022). A review of predictive policing from the perspective of fairness. Artificial Intelligence and Law, 30, 1–17.

Angwin, J., Larson, J., Mattu, S., & Kirchner, L. (2022). Machine bias. In Ethics of Data and Analytics (pp. 254–264). Auerbach Publications.

Bellamy, R. K. E., Dey, K., Hind, M., Hoffman, S. C., Houde, S., Kannan, K., Lohia, P., et al. (2018). AI Fairness 360: An extensible toolkit for detecting, understanding, and mitigating unwanted algorithmic bias. arXiv preprint arXiv:1810.01943.

Bird, S., Dudík, M., Edgar, R., Horn, B., Lutz, R., Milan, V., Sameki, M., et al. (2020). Fairlearn: A toolkit for assessing and improving fairness in AI. Microsoft Research. MSR-TR-2020-32. https://www.microsoft.com/en-us/research/publication/fairlearn-a-toolkit-for-assessing-and-improving-fairness-in-ai/

Buolamwini, J., & Gebru, T. (2018). Gender shades: Intersectional accuracy disparities in commercial gender classification. In Conference on Fairness, Accountability and Transparency (pp. 77–91). PMLR.

Cheng, L., Varshney, K. R., & Liu, H. (2021). Socially responsible AI algorithms: Issues, purposes, and challenges. Journal of Artificial Intelligence Research, 71, 1137–1181.

Holstein, K., Vaughan, J. W., Daumé III, H., Dudik, M., & Wallach, H. (2019). Improving fairness in machine learning systems: What do industry practitioners need? In Proceedings of the 2019 CHI Conference on Human Factors in Computing Systems (pp. 1–16).

IBM (2022). Everyday Ethics for Artificial Intelligence. https://www.ibm.com/watson/assets/duo/pdf/everydayethics.pdf.

Jobin, A., Ienca, M., & Vayena, E. (2019). The global landscape of AI ethics guidelines. Nature Machine Intelligence, 1(9), 389–399.

Noble, S. U. (2018). Algorithms of Oppression. New York University Press.

Ribeiro, M. T., Singh, S., & Guestrin, C. (2016). "Why should I trust you?" Explaining the predictions of any classifier. In Proceedings of the 22nd ACM SIGKDD International Conference on Knowledge Discovery and Data Mining (pp. 1135–1144).

Richardson, B., Garcia-Gathright, J., Way, S. F., Thom, J., & Cramer, H. (2021). Towards fairness in practice: A practitioner-oriented rubric for evaluating Fair ML Toolkits. Proceedings of the 2021 CHI Conference on Human Factors in Computing Systems (pp. 1–13).

Richardson, B., & Gilbert, J. E. (2021). A framework for fairness: A systematic review of existing fair AI solutions. arXiv preprint arXiv:2112.05700.

Saleiro, P., Kuester, B., Hinkson, L., London, J., Stevens, A., Anisfeld, A., Rodolfa, K. T., et al. (2018). Aequitas: A bias and fairness audit toolkit. arXiv preprint arXiv:1811.05577.

Sweeney, L. (2013). Discrimination in online ad delivery. Communications of the ACM, 56(5), 44-54.

The National Archives. (2010). Equality Act 2010, c. 15. Retrieved from https://www.legislation.gov.uk/ukpga/2010/15/contents

Veale, M., Van Kleek, M., & Binns, R. (2018). Fairness and accountability design needs for algorithmic support in high-stakes public sector decision-making. In Proceedings of the 2018 CHI Conference on Human Factors in Computing Systems (pp. 1–14).

Verma, S., & Rubin, J. (2018). Fairness definitions explained. In Proceedings of the International Workshop on Software Fairness (pp. 1–7).

End of Chapter Questions

1. What are the three categories of fairness measures provided by Verma and Rubin?

2. What are the three types of biases adopted for a "diverse range of biases" criterion used in the experiments?

3. Which type of machine learning task is each toolkit capable of?

4. What is the difference between bias detection and bias mitigation in the context of the toolkits?

5. Table 4.1 gives the number of TP, TN, FP, and FN for a reference group and target group. Please check all the disparities are preserved for {**sex:'Male', race:'African-American', Age:'Greater than 45'**}. Any disparity is passed if the metric for the group over the metric for the reference group is less than a threshold. The threshold is 0.5 and the reference group is {**sex:'Male', race:'Caucasian', Age:'Less than 25'**}.

$$1 - \tau \le FDR \text{ disparity black} \le \frac{1}{(1-\tau)}$$

$$FDR \text{ disparity black} = \frac{FDR \text{ black}}{FDR \text{ reference group}}$$

   a. FDR disparity
   b. FPR disparity
   c. FOR disparity
   d. FNR disparity
   e. TPR disparity
   f. TNR disparity

6. List four findings from Figure 4.4 that summarize the plot.

7. Figure 4.5 shows that all of the age groups pass the fairness threshold values when individuals between 25 and 45 are the reference group. List your main findings across these groups.

8. Results from bias mitigation with Fairlearn are shown in Figures 4.6 to 4.8. List four findings you can infer from the graphs.

Metric	Reference Group {sex: 'male', race: 'Caucasian', age: 'Less than 25'}	Target Group {sex: 'male', race: 'African-American', age: 'Greater than 45'}
TP	11	13
FP	4	9
FN	25	26
TN	37	71

TABLE 4.1 Reference and Target Group Metrics

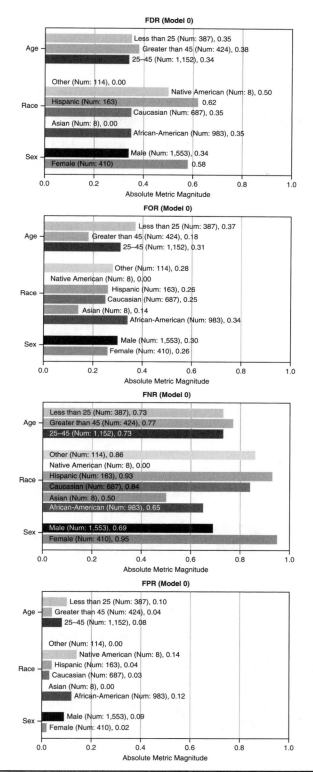

Figure 4.4 Intersectional analysis of FDR, FOR, FNR, and FPR across race, sex, and age.

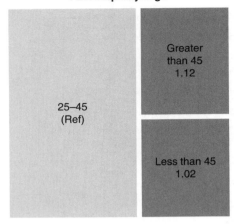

FDR Disparity: Age

25–45
(Ref)

Greater
than 45
1.12

Less than 45
1.02

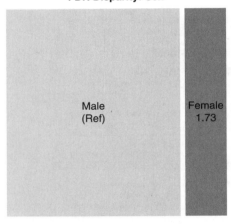

FDR Disparity: Sex

Male
(Ref)

Female
1.73

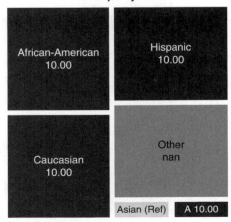

FDR Disparity: Race

African-American
10.00

Hispanic
10.00

Caucasian
10.00

Other
nan

Asian (Ref)   A 10.00

Not labeled above:
A: Native American 10.00

FIGURE 4.5 False discovery rate (FDR) disparity based on age, sex, and race.

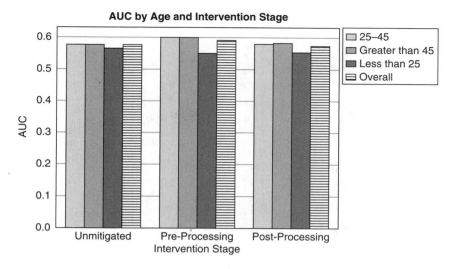

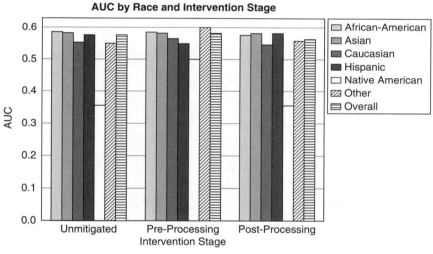

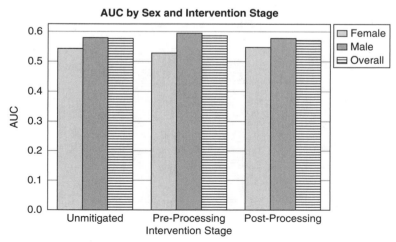

FIGURE 4.6 Bias mitigation effects on area under the curve (AUC) using age, race, and sex as sensitive attributes.

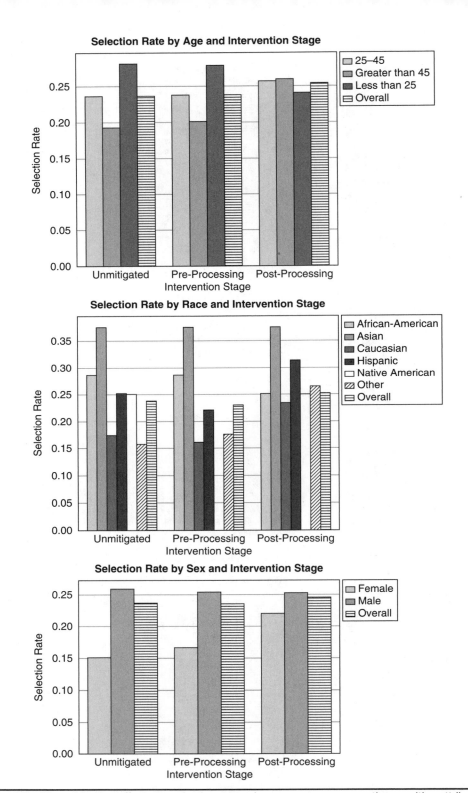

FIGURE 4.7 Bias mitigation effects on selection rate using age, race, or sex as the sensitive attribute.

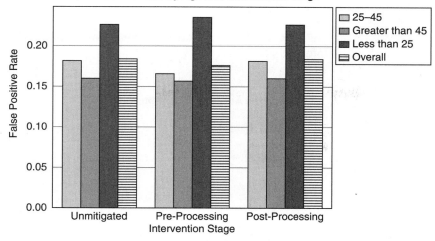

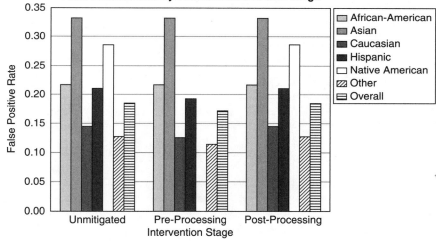

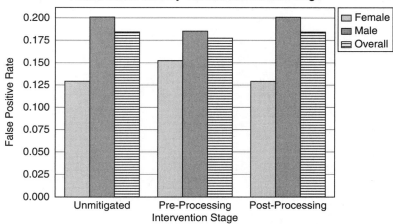

Figure 4.8 Bias mitigation effects for false positive rate (FPR) using age, race, and sex as sensitive attributes.

9. Results from bias mitigation with AIF360 are shown in Figures 4.9 to 4.11. List four findings you can infer from the graphs.

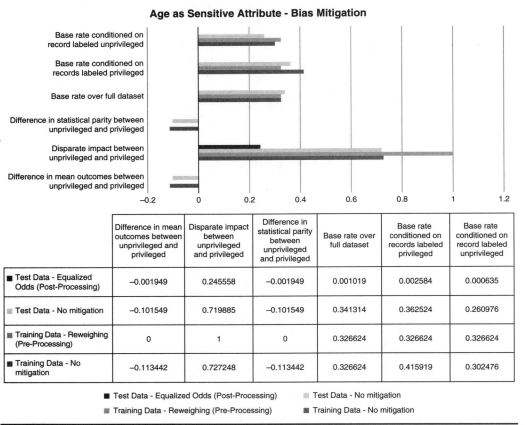

	Difference in mean outcomes between unprivileged and privileged	Disparate impact between unprivileged and privileged	Difference in statistical parity between unprivileged and privileged	Base rate over full dataset	Base rate conditioned on records labeled privileged	Base rate conditioned on record labeled unprivileged
■ Test Data - Equalized Odds (Post-Processing)	−0.001949	0.245558	−0.001949	0.001019	0.002584	0.000635
▨ Test Data - No mitigation	−0.101549	0.719885	−0.101549	0.341314	0.362524	0.260976
▨ Training Data - Reweighing (Pre-Processing)	0	1	0	0.326624	0.326624	0.326624
■ Training Data - No mitigation	−0.113442	0.727248	−0.113442	0.326624	0.415919	0.302476

■ Test Data - Equalized Odds (Post-Processing)   ▨ Test Data - No mitigation
▨ Training Data - Reweighing (Pre-Processing)   ■ Training Data - No mitigation

FIGURE 4.9   Bias mitigation effects using age as a sensitive attribute.

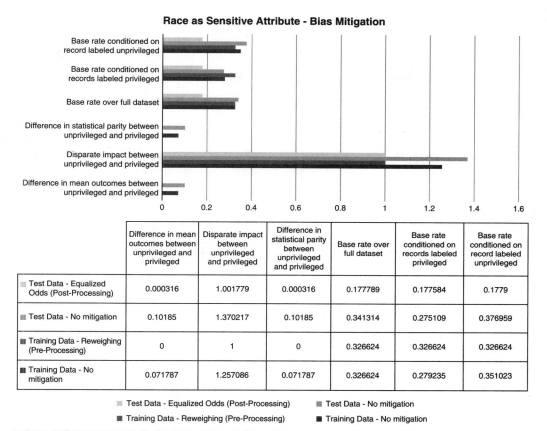

	Difference in mean outcomes between unprivileged and privileged	Disparate impact between unprivileged and privileged	Difference in statistical parity between unprivileged and privileged	Base rate over full dataset	Base rate conditioned on records labeled privileged	Base rate conditioned on record labeled unprivileged
Test Data - Equalized Odds (Post-Processing)	0.000316	1.001779	0.000316	0.177789	0.177584	0.1779
Test Data - No mitigation	0.10185	1.370217	0.10185	0.341314	0.275109	0.376959
Training Data - Reweighing (Pre-Processing)	0	1	0	0.326624	0.326624	0.326624
Training Data - No mitigation	0.071787	1.257086	0.071787	0.326624	0.279235	0.351023

Test Data - Equalized Odds (Post-Processing)   Test Data - No mitigation
Training Data - Reweighing (Pre-Processing)   Training Data - No mitigation

FIGURE 4.10   Bias mitigation effects using race as a sensitive attribute.

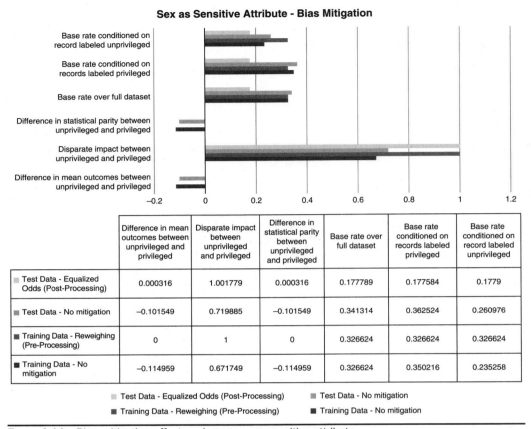

Sex as Sensitive Attribute - Bias Mitigation

	Difference in mean outcomes between unprivileged and privileged	Disparate impact between unprivileged and privileged	Difference in statistical parity between unprivileged and privileged	Base rate over full dataset	Base rate conditioned on records labeled privileged	Base rate conditioned on record labeled unprivileged
▨ Test Data - Equalized Odds (Post-Processing)	0.000316	1.001779	0.000316	0.177789	0.177584	0.1779
▪ Test Data - No mitigation	−0.101549	0.719885	−0.101549	0.341314	0.362524	0.260976
▪ Training Data - Reweighing (Pre-Processing)	0	1	0	0.326624	0.326624	0.326624
▪ Training Data - No mitigation	−0.114959	0.671749	−0.114959	0.326624	0.350216	0.235258

▨ Test Data - Equalized Odds (Post-Processing)   ▪ Test Data - No mitigation
▪ Training Data - Reweighing (Pre-Processing)   ▪ Training Data - No mitigation

FIGURE 4.11   Bias mitigation effects using sex as a sensitive attribute.

10. Think of a real-world application where you might use these fairness toolkits. What is the application? What are the sensitive attributes? Are there any fairness checks you would like to perform that are not included in these toolkits?

11. This chapter discussed datasets made from the records of individual humans. These datasets may contain protected attributes, like race or sex, or proxies for them, like neighborhoods. Now think about datasets that are made up of nonhuman data (for example, air pollution prediction or plant disease classification, though there are many more). How do you think these fairness metrics would apply to datasets like this? Do you think they are necessary for nonhuman data?

CHAPTER **5**

Bias Mitigation in Hate Speech Detection

Zahraa Al Sahili

Queen Mary University of London

Question: How can you mitigate bias in hate speech detection systems?

Learning Objectives

Upon completion of this chapter, the student should be able to

- Provide an overview of machine learning–based hate speech detection systems
- Explore bias mitigation methods: transfer learning, multitask learning, and adversarial methods
- Implement hate speech detection systems and mitigate various biases using multitask learning

Chapter Overview

In this chapter, we set the context and provide basic bias mitigation methods in machine learning when applied to hate speech detection systems.

5.1   Introduction

In recent years, the issue of hate speech has garnered increased attention, prompting the development of automated systems to detect and prevent its spread online. However, detecting hate speech remains a challenging task due to its subjective nature, which is highly dependent on societal perspectives such as a person's identity or dialect. Furthermore, hate speech detection systems, like any other natural language processing (NLP) systems, are susceptible to biases that may harm historically excluded groups. To address these ethical concerns, it is essential to employ bias mitigation methods before deploying these models. In this chapter, we will examine different approaches to bias mitigation in hate speech detection systems, highlighting the importance of ethical considerations in the development and implementation of these technologies.

In the realm of hate speech detection, ensuring unbiased and fair outcomes is of utmost importance. Machine learning models play a crucial role in analyzing and identifying harmful content at scale, but they are not immune to biases that can undermine their effectiveness. As practitioners, it is our responsibility to address these biases head-on and implement strategies that mitigate their impact.

One prevalent approach to combating bias in hate speech detection systems is transfer learning. By leveraging preexisting models trained on vast amounts of general language data, we can initialize our models with rich linguistic knowledge. However, it is vital to consider the potential transfer of biases present in the training data. A thorough understanding of the biases encoded in the source model and careful fine-tuning can help us adapt the model to detect hate speech while minimizing the amplification of existing biases.

Multitask learning is another valuable technique for bias mitigation. By simultaneously training models on multiple related tasks, such as sentiment analysis or topic classification, we can encourage the model to learn more generalized representations of language. This broader perspective can help counteract specific biases inherent in hate speech detection, making the model more robust and less susceptible to biased judgments.

Adversarial methods provide yet another avenue for addressing bias. These techniques involve training two competing models—a classifier and an adversarial discriminator—to continuously challenge and improve each other. The classifier strives to accurately detect hate speech, while the adversarial discriminator attempts to identify and exploit any underlying biases in the classifier's decisions. This iterative process helps uncover and neutralize biases, leading to fairer and more balanced hate speech detection systems.

In this chapter, we will explore these bias mitigation methods—transfer learning, multitask learning, and adversarial methods—in depth. We will examine the theoretical foundations of each approach and provide practical guidance on their implementation. Through a combination of hands-on exercises, case studies, and real-world examples, we will equip you with the knowledge and skills to effectively mitigate biases in hate speech detection systems.

By employing these methods, we can move closer to creating hate speech detection systems that are not only accurate but also fair, respectful, and reflective of the diverse perspectives found in society. Bias mitigation is an ongoing endeavor that demands continuous learning and adaptation, and this chapter will empower you to take meaningful steps toward building more inclusive and less biased machine learning models for hate speech detection.

5.2   Background

Hate speech detection systems in machine learning are designed to identify and mitigate offensive, discriminatory, or harmful content in digital communication. These systems leverage advanced NLP techniques, such as deep learning models, to analyze text and classify it as hate speech or nonhate speech. By training on large datasets containing labeled examples, these systems learn to recognize patterns and linguistic cues indicative of hate speech. Key challenges include addressing the dynamic nature of language, handling the subtleties of context, and ensuring that the system remains unbiased across diverse demographics. Hate speech detection systems play a crucial role in promoting online safety, fostering inclusivity, and maintaining respectful digital spaces. As artificial intelligence (AI) and NLP technologies evolve, the continuous improvement of these systems becomes essential to combatting the proliferation of hate speech in our interconnected world.

Now we will define concepts related to machine learning and hate speech detection methods that are essential for bias mitigation understanding.

Transfer learning is a vital technique in hate speech detection systems, involving the utilization of pretrained models as a foundation for solving new hate speech detection tasks. Instead of starting from scratch, transfer learning employs a pretrained model and fine-tunes it for the specific hate speech detection task. This approach is particularly advantageous when dealing with limited hate speech data, enabling faster and more accurate learning.

Multitask learning is a crucial strategy for hate speech detection systems, enabling a single model to simultaneously learn multiple interconnected hate speech detection tasks. This technique is valuable when there are interdependencies between hate speech detection tasks or when individual task data are scarce. By leveraging shared knowledge across these tasks, multitask learning significantly enhances the accuracy of detecting hate speech while improving overall model efficiency.

Generative adversarial networks (GANs) are a neural network architecture employed in hate speech detection systems to generate synthetic hate speech data resembling the training data. GANs consist of two key components, a generator and a discriminator, trained in a game-theoretic framework. The generator aims to create hate speech data that can deceive the discriminator, which, in turn, strives to distinguish between real and fake hate speech data. Through this adversarial training, GANs become adept at generating new hate speech data that closely resemble the original training data, facilitating more comprehensive detection.

Recurrent neural networks (RNNs) serve as a foundational neural network architecture for processing sequential hate speech data. RNNs incorporate loops in their structure to maintain information continuity over time, making them highly suitable for hate speech detection in contexts like NLP, speech recognition, and time series prediction.

Long short-term memory (LSTM) networks, a specific type of RNN, play a crucial role in hate speech detection systems by addressing the vanishing gradient problem commonly encountered during the training of deep neural networks. LSTMs employ memory cells and gates to selectively retain or forget information over time, enabling more effective processing of lengthy sequences of hate speech data compared to traditional RNNs.

In addition to the advantages of these concepts in hate speech detection systems, they are also essential in mitigating bias, which we will discuss in the next sections.

5.2.1 Case Study of Hate Speech Detection

Now let's walk through a simplified tutorial-style case study on detecting and removing hate speech using a fictional dataset. Please note that this is a simplified example for educational purposes; real-world systems may have more complexity and use advanced techniques. Our steps will include data preprocessing, model training, and hate speech detection. Use the Python source code attached with the book chapter while following up with the case study.

Step 1: Dataset Preparation

Let's create a small fictional dataset with text samples labeled "hate speech" or "not hate speech" (see Figure 5.1). We'll use a simple comma-separated value (CSV) format with a "text" column and a "label" column.

We'll preprocess the text data to convert it into a numerical format that can be fed into a machine learning model. We'll tokenize the text; remove stop words; and convert it to lowercase, using Python, pandas, and sklearn.

Now load the dataset, split data into features (X) and labels (y), then split the data into training and testing sets, and finally tokenize and vectorize the text data.

Step 2: Model Training

We'll use a simple classifier, such as a logistic regression, for hate speech detection. We'll train the model on the preprocessed training data. Train a logistic regression model, make predictions on the test set, and then evaluate the model.

Step 3: Hate Speech Detection

Now, use the trained model to detect hate speech in new text samples. If hate speech is detected, we'll remove the corresponding content. Preprocess the input text and then predict using the trained model.

Example usage:

```
detect_and_remove_hate_speech("You're amazing, keep it up!") # Not hate
speech
detect_and_remove_hate_speech("I hate your views!") # Hate speech
(removed)
```

In this fictional example, we created a small dataset, preprocessed the data, trained a simple logistic regression model, and demonstrated how to detect and remove hate speech using the trained model. In a real-world scenario, a more sophisticated model, larger dataset, and robust text processing pipeline would be used to handle hate speech detection at scale. Additionally, the process of removal would likely involve moderation or filtering mechanisms depending on the platform or context.

5.2.2   Section Summary

In this section, we provided an overview of some basic concepts in hate speech detection systems and implemented a naïve hate speech detection system based on logistic regression.

```
text,label
"Get out of here, you're not welcome!",hate
"I disagree with your opinion.",not_hate
"Spread love and positivity!",not_hate
"Hateful words have no place.",not_hate
"Go back to your country!",hate
```

FIGURE 5.1   Hate speech dataset.

5.3   Bias in Hate Speech Detection Systems

As part of NLP systems, hate speech detection models are vulnerable to many types of biases. Such systems face plenty of challenges due to atypical changes in spelling and grammar, the absence of a uniform definition of hate speech, and the presence of unintended identity biases (Mozafari et al., 2020). For example, consider the offensive tweet "No for refugees" that will be flagged as hate speech and reported, as our machine learning systems are trained to report any tweet expressing hate speech, including the hate of refugees. However, the model can tend to classify any tweet containing a refugee identity like "Syrian" as hate speech, so it will be reporting a neutral tweet like "I am a Syrian refugee" as offensive. This is mainly because the keywords "refugee" and "Syrian" are repeated in hate speech tweets, causing the hate speech systems to flag tweets containing these keywords even if they are not hate speech. To address such issues, various bias mitigation techniques have been developed, including fine-tuning, multitask learning, and adversarial training, which aim to mitigate the impact of single or multiple biases in hate speech detection models. In the next three sections we will discuss in detail the popular bias mitigation methods in hate speech detection systems.

5.4   Bias Mitigation in Hate Speech Detection Using Transfer Learning

In the realm of hate speech detection, transfer learning has emerged as a crucial technique to leverage the knowledge captured by pretrained models while addressing the challenges posed by bias. Transfer learning enables hate speech detection systems to benefit from models trained on vast amounts of data, even if that data don't precisely match the target domain. This section explores the application of transfer learning to mitigate bias in hate speech detection.

- **Leveraging pretrained models**: Transfer learning involves utilizing pretrained models, such as those trained on large-scale language tasks, as a foundation for hate speech detection. By leveraging the knowledge captured by these models, we can bootstrap the hate speech detection process, especially when the available hate speech data are limited.

- **Fine-tuning for bias reduction**: Fine-tuning, a key aspect of transfer learning, allows us to adapt pretrained models to the specific hate speech detection task. In this context, fine-tuning includes strategies to reduce bias by carefully considering the distribution of labels and the potential sources of bias in the data. Techniques such as reweighting or resampling can be employed to mitigate biases.

After the successes of transfer learning in improving performance for models with scare data, transfer learning was expanded to mitigate bias in hate speech systems. The idea is simple: we fine-tune the neural network that has biased data on pretrained models that were trained on debiased data (see Figure 5.2). However, this approach is restricted to debiased pretrained model availability, mostly in popular languages like English. In addition, the bias types mitigated are the types only debiased in the pretrained models.

For example, Jin et al. used transfer learning for bias mitigation in hate speech classification using downstream fine-tuning (Jin et al., 2020). The debiased upstream

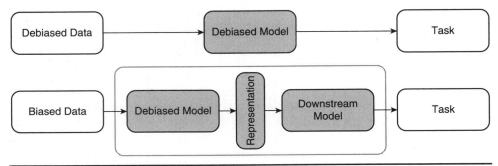

FIGURE 5.2   Transfer learning.

models are less biased upon transfer to downstream models when used on new tasks that share similarity with the upstream model's task but in a new domain. Another transfer learning approach was proposed by Li to deal with bias in the case of class imbalance (Li, 2021). The method is based on the adversarial discriminative domain adaptation (ADDA), where first a classifier is trained on the source data and then the feature extractor is learned for the target domain by acting like a generator that is trying to fool a discriminator.

5.4.1   Case Study of Transfer Learning

Addressing bias in hate speech detection using transfer learning involves leveraging pretrained models and fine-tuning them to reduce bias. In this example, we'll use a fictional dataset and demonstrate how to fine-tune a pretrained model to detect hate speech while mitigating bias. Use the Python source code attached with the book chapter while following up with the case study.

Step 1: Dataset Preparation
Let's create a fictional dataset that includes text samples labeled as "hate speech" or "not hate speech," along with demographic attributes that might indicate bias (Figure 5.3).

Step 2: Data Preprocessing
We'll preprocess the text data and encode the demographic attributes to numerical format for fine-tuning using Python and the packages pandas, sklearn, pytorch, and transformers.

Now load the dataset, encode the demographic attributes, split the data into training and testing sets, and finally tokenize the text data.

```
text,label,gender,age
"Get out of here, you're not welcome!",hate,male,30
"I disagree with your opinion.",not_hate,female,25
"Spread love and positivity!",not_hate,male,40
"Hateful words have no place.",not_hate,female,30
"Go back to your country!",hate,male,35
```

FIGURE 5.3   Dataset preparation.

Step 3: Model Fine-Tuning

We'll use a pretrained bidirectional encoder representations from transformers (BERT) model for fine-tuning, and we'll take into account the demographic attributes to reduce potential bias.

Step 4: Bias Mitigation

To mitigate bias, we can analyze the model's performance across different demographic groups and take corrective measures if necessary. We can also implement fairness-aware training techniques and postprocessing methods to ensure fair and unbiased hate speech detection.

In this fictional example, we demonstrated how to fine-tune a pretrained BERT model for hate speech detection while considering demographic attributes to reduce bias. In practice, a more comprehensive analysis of bias and fairness would be necessary, and a real-world system would involve a larger, more diverse dataset and more sophisticated techniques to address bias effectively.

5.4.2   Section Summary

In this section, we discussed the transfer learning method for reducing bias in hate speech detection.

5.5   Bias Mitigation in Hate Speech Detection Using Transfer Learning

Multitask learning, where a single model learns multiple related tasks simultaneously, offers a powerful paradigm to enhance fairness in hate speech detection. By jointly considering the detection of hate speech and demographic attributes, we can develop models that are more sensitive to potential sources of bias.

- **Simultaneous hate speech detection and demographic prediction**: In multitask learning, hate speech detection is coupled with the prediction of demographic attributes such as gender, age, or ethnicity. This approach allows the model to recognize correlations between demographic features and hate speech, thus providing insights into potential bias.

- **Shared knowledge**: Multitask learning enables the model to leverage shared knowledge across tasks. When demographic prediction is one of the tasks, the model can use this information to enhance its understanding of the data and make more informed predictions about hate speech, while being aware of potential biases associated with the demographic attributes.

- **Bias-aware training**: Multitask learning encourages the model to be sensitive to bias by incorporating demographic attributes into the learning process. Techniques like adversarial training can be applied to explicitly minimize the effect of bias in the model's predictions, leading to a more fair and unbiased hate speech detection system.

Another efficient method to mitigate bias in hate speech is multitask learning. Adding another auxiliary task while training the hate detection system, like identity detection, results in bias mitigation for various identities. This approach can be applied

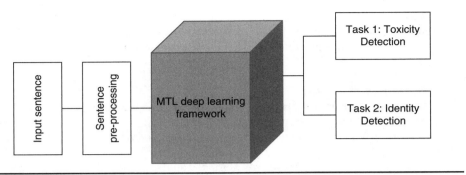

Figure 5.4   Multitask learning.

for any language, any bias type, and more than one bias type through adding more than one auxiliary task. However, it is limited to the availability of identity labels for the data available. The mitigation process is illustrated in Figure 5.4, where after pre-processing hate speech sentences, we select a deep learning framework that has two outputs, toxicity detection and identity detection, where identity can be of any type: gender, religion, racial, and so on. During training, multitask learning is applied where part of the models' weights are shared between the tasks yielding bias mitigation.

For example, to mitigate unintended identity bias, including gender, religion, race, and mental health status, the state of the art of Vaidya et al. applied an attention-based multitask learning approach, where the identity classification task is accompanied by the toxicity detection task (Vaidya et al., 2019). Attention is added to identity classification for decreasing identity bias (Vaidya et al., 2019). On average, the model was able to predict 10% less toxicity for nontoxic comments in the studied underrepresented identities (Vaidya et al., 2019). Multitask learning was also applied by Faal et al. using domain adaptation multitask learning to mitigate identity bias in toxic language detection (Faal et al., 2021). Pretraining on the BERT model was followed by an anti-curriculum multitask learning framework (Faal et al., 2021). First, identity detection tasks including religion, gender, race, and disability are trained, and then after two epochs, the toxicity detection task is progressed (Faal et al., 2021). This resulted in a better area under the curve (AUC) score for the four identities compared to the fine-tuned model (Faal et al., 2021). The adaptive-BERT multitask model achieved AUC accuracies of 0.959, 0.958, 0.948, 0.952, and 0.954 for gender, religion, race, sexual orientation, and disability, respectively, compared to 0.94, 0.946, 0.933, 0.926, and 0.944 AUC scores for the same identities in the BERT fine-tuned baseline model (Faal et al., 2021). Thus, the proposed approach efficiently mitigated unintended bias for the desired minorities (Faal et al., 2021).

5.5.1   Case Study on Multitask Learning

To reduce bias in hate speech detection using multitask learning, we'll leverage the demographic attributes as auxiliary tasks to help the model learn to detect hate speech while considering potential bias. We'll use a shared model architecture that handles both hate speech detection and the prediction of demographic attributes simultaneously. Let's walk through the steps. Use the Python source code attached with the book chapter while following up with the case study.

Step 1: Dataset Preparation

We'll use the same fictional dataset with text samples, labels, gender, and age as before.

Step 2: Data Preprocessing

We'll preprocess the text data and encode the demographic attributes for multitask learning. Load the dataset, split the data into features (text and demographics) and labels, encode the demographic attributes, and then split the data into training and testing sets.

Step 3: Model Architecture for Multitask Learning

We'll define a shared model architecture that handles both hate speech detection and demographic attribute prediction. Define the shared model for multitask learning, then load the pretrained BERT model for sequence classification, and then define the optimizer. After that create a DataLoader for the training data, go through a multitask fine-tuning loop, and finally valuate the multitask model.

In this example, we demonstrated how to use a shared model architecture for multitask learning, where the model simultaneously predicts hate speech labels and demographic attributes. The shared architecture allows the model to consider demographic attributes as auxiliary tasks, potentially helping to reduce bias in hate speech detection. In practice, comprehensive bias analysis and fairness-aware techniques would be essential to ensure effective bias reduction.

5.5.2   Section Summary

In this section we investigated multitask learning as a bias mitigation technique.

5.6   Adversarial Methods for Bias Reduction in Hate Speech Detection

Adversarial methods offer a unique approach to bias reduction by explicitly addressing sources of bias through a game-theoretic framework. By introducing an adversarial component, we can encourage the model to reduce bias in its predictions, leading to more equitable hate speech detection systems.

- **Adversarial components**: Adversarial methods introduce an adversarial network that attempts to counteract biases present in the model's predictions based on demographic attributes. This adversarial component plays a critical role in minimizing the effects of bias while optimizing the main task of hate speech detection.

- **Minimizing bias through adversarial training**: Adversarial training encourages the model to make predictions that are robust against biases. The adversarial component learns to balance the impact of demographic attributes, ensuring that the model's decisions are fair and unbiased across different demographic groups.

- **Fairness-aware evaluation**: Adversarial methods promote fairness-aware evaluation, allowing us to assess the model's performance across demographic attributes. We can measure the model's bias reduction effectiveness, ensuring that the hate speech detection system remains unbiased and equitable, regardless of the underlying demographic characteristics.

For example, Madras et al. proposed a fair-generating representation algorithm using adversarial learning (Madras et al., 2018). The approach aimed to minimize the expectedness of maintained features from the input data while optimizing the classifier by increasing its classification accuracy. Another adversarial method to maximize the predictor's ability in outputting correct predictions is proposed by Zhang et al. using two tasks: analogy completion and classification. It achieved higher accuracy in the word embedding task (Zhang et al., 2018). In addition, adversarial model methods effectively reduced the racial bias on the FDCL18 and Brod16 datasets while performing poorly on the DWMW17 dataset (Xia et al., 2020). Failing to mitigate bias in the DWMW17 dataset was caused by an extreme imbalance in the tweet's labels, as 97% of the AEE dialects were labeled as toxic tweets (Xia et al., 2020). The adversarial proposed approach can readily be applied to other bias types such as gender and religious biases (Xia et al., 2020). However, adversarial methods are inefficient in detecting bias and predicting toxicity in an imbalanced dataset. Moreover, Morzhov et al. aimed to build two models that can reduce the unintended bias and detect toxicity attributes effectively (Morzhov, 2020). The first model (bi-GRU-LSTM) is based on a combination of gated recurrent units (GRUs) and LSTM models (Morzhov, 2020). The second model was approximately similar to the first one but with an attention mechanism where different words will be assigned different weights based on their toxicity. It used the BERT method to get word embeddings, and the models were built after two stages of data preprocessing techniques (Morzhov, 2020). The methods showed a drop in toxicity score in comparison with the old perspective application programming interface (API) model when tested on complex sentences.

5.6.1   Case Study on Adversarial Training

Adversarial training is a powerful approach for improving the robustness and fairness of machine learning models, especially in the context of bias reduction. In this example, we'll use adversarial training to reduce bias in hate speech detection while considering demographic attributes as potential sources of bias. We'll create an adversarial network to counteract potential biases in the model's predictions based on the demographic attributes. Let's walk through the steps. Use the Python source code attached with the book chapter while following up with the case study.

Step 1: Dataset Preparation
We'll use the same fictional dataset with text samples, labels, gender, and age as before.

Step 2: Data Preprocessing
We'll preprocess the text data and encode the demographic attributes for adversarial training, Load the dataset, split the data into features (text and demographics) and labels, and then encode the demographic attributes, while splitting the data into training and testing sets.

Step 3: Model Architecture with Adversarial Component
We'll define a model architecture that includes an adversarial component aiming to mitigate bias based on the demographic attributes. Define the model with an adversarial component, forward-pass through the BERT model, then forward-pass through the demographic predictor, then forward-pass through the adversarial component, and finally define the optimizer.

Step 4: Adversarial Training

We'll train the model using an adversarial loss that encourages the adversarial component to be uncertain about the demographic attributes while accurately predicting hate speech labels. First go through the adversarial training loop and then evaluate the adversarial model.

In this example, we demonstrated how to use an adversarial component to mitigate bias in a hate speech detection model while considering demographic attributes as potential sources of bias. The adversarial training encourages the model to make unbiased predictions based on the demographic attributes. In practice, more advanced fairness-aware techniques and extensive evaluation of bias reduction would be necessary for a comprehensive solution.

5.6.2 Section Summary

This section focuses on bias reduction through adversarial training.

5.7 Benefits and Pitfalls

Addressing bias in hate speech detection is crucial for developing equitable and effective systems. Different techniques, such as transfer learning, multitask learning, and adversarial methods, offer unique approaches to mitigate bias. Each approach has made notable advancements but also faces specific pitfalls that must be considered.

5.7.1 Transfer Learning

Advancements

- **Robust pretrained models**: Recent pretrained models (e.g., BERT) capture nuanced language patterns, aiding in the identification of subtle forms of hate speech while mitigating bias across diverse language styles.
- **Domain adaptation techniques**: Advances in domain adaptation methods enable models to adapt to different data distributions, which is valuable for addressing bias arising from varying demographics or platforms.

Pitfalls

- **Data bias**: Transfer learning heavily relies on the quality of pretraining data. Biases in the pretraining data may propagate to fine-tuned models, making addressing data biases a significant challenge.
- **Bias amplification**: Fine-tuning on a limited hate speech dataset may unintentionally amplify biases present in that data, emphasizing the need for diverse fine-tuning datasets.

5.7.2 Multitask Learning

Advancements

- **Enhanced bias understanding**: Multitask learning offers insights into the relationship between demographic attributes and hate speech, helping guide bias reduction strategies.

- **Fairness-aware objectives**: Advanced fairness-aware objectives can be integrated into multitask learning, promoting balanced predictions across demographic groups.

Pitfalls
- **Demographic noise**: Noisy or inaccurate labels in demographic attributes can negatively impact model performance, leading to biased predictions and requiring careful data preprocessing.
- **Model complexity**: Balancing the objectives of hate speech detection and demographic prediction requires fine-tuning to prevent overfitting or suboptimal performance.

5.7.3   Adversarial Methods

Advancements
- **Explicit bias mitigation**: Adversarial components explicitly target bias reduction, introducing a separate network to counteract the effects of demographic attributes on predictions.
- **Fairness evaluation metrics**: Adversarial methods encourage the development of novel fairness evaluation metrics, providing a clearer understanding of bias reduction effectiveness.

Pitfalls
- **Adversarial training complexity**: Adversarial training can be challenging to optimize and may require careful hyperparameter tuning to prevent the adversarial component from dominating the training process.
- **Adversarial attacks**: Ensuring the adversarial component's robustness against adversarial attacks is essential to maintaining bias reduction effectiveness.

Each technique offers valuable tools for bias mitigation in hate speech detection, but careful consideration of their specific advances and pitfalls is essential. Incorporating these techniques and addressing their challenges can lead to more fair, more robust, and less biased hate speech detection systems in today's diverse digital landscape.

5.7.4   Section Summary
In this section we compared common bias mitigation algorithms in hate speech detection systems.

5.8   Other Methods
Other methods exist to mitigate bias in hate speech detection systems. Zhao et al. (2018) found that three systems trained on the same dataset where females are less represented are subject to gender bias. Thus, bias source is the gender imbalance in the datasets and word embeddings (Zhao et al., 2018). To address this, two strategies were introduced: gender-swapping to obtain a gender-balanced dataset and replacing word embeddings with debiased vectors (Zhao et al., 2018).

On the other side, Sap et al. investigated racial bias by finding correlations between offensive labeling and the dialect of African American English (AAE) in famous datasets annotated with toxic language (Sap et al., 2019). In both datasets, a strong Pearson correlation (r) was shown between AAE dialect and different categories of hate speech ($r = 0.24$ in "offensive" label and $r = 0.35$ in "abusive" label), which are proof of bias due to the dialect in such datasets (Sap et al., 2019). To mitigate annotator bias in terms of both dialect and race priming, stress on the tweet dialect was used in the annotation of data (Sap et al., 2019).

In addition, Park et al. studied the efficiency of three methods to reduce the gender bias: debiasing word embeddings, gender swap data augmentation, and fine-tuning with a larger corpus (Park et al., 2018). Applied separately, debiasing word embedding was the least effective method, while gender swap data augmentation was the most effective (Park et al., 2018). The optimum technique was using the mitigation approaches together, where gender bias was decreased between 90% and 98% when applied on multiple datasets (Park et al., 2018).

5.8.1   Section Summary

This section mentions briefly bias mitigation approaches other than transfer learning, multitask learning, and adversarial training.

5.9   Hands-on Exercise

Now we go through a hands-on exercise using bias mitigation in a hate speech detection system using multitask learning. Follow along with the hands-on exercise in the Jupyter notebook that accompanies the book.

5.9.1   Model

The model is composed of an embedding layer, one single-layer bi-LSTM encoder, a single LSTM shared layer, and a specific LSTM layer for each task followed by a specific fully connected output layer (Figure 5.5).

5.9.2   Dataset

For this exercise we will select the multilingual and multi-aspect (MLMA) dataset. The original dataset has five annotation aspects describing (1) the directness, (2) the hostility type, (3) the target attribute, (4) the target group, and (5) the sentiment of the annotator. Each of the labeled classes can be used as a classification task. These different labels were designed by Ousidhoum et al. (2019) to facilitate the study of correlation between explicitness of the tweet, the type of hostility it conveys, its target attribute, the group it dehumanizes, how different people react to it, and the performance of multitask learning on the five tasks for three different languages: Arabic, English, and French. For the hands-on part, we will work on the English part of the dataset.

We will use Python and popular frameworks such as pandas, matplotlib, sklearn, TensorFlow, keras, and NLTK.

5.9.3   Data Preprocessing

1. First, we preprocessed the sentences in our dataset. The preprocessing procedure consists of anonymization of the tweets by removing @user and @url if applicable.

2. Remove the stop words.

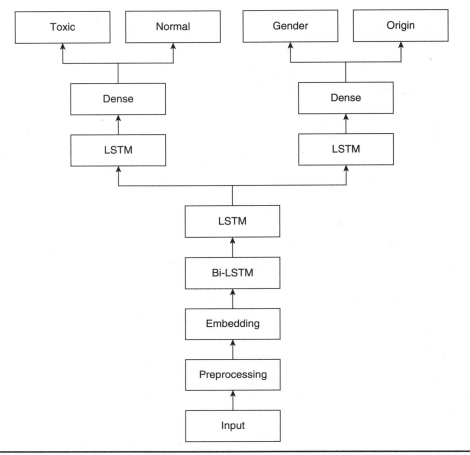

FIGURE 5.5 Multitask learning LSTM-based model.

3. Remove all punctuation and nonalphabetical characters, such as emojis.

4. Remove English characters and numbers.

5. Normalize the words by removing diacritics and letters that stand for pronouns.

5.9.4 Hate Speech Classification

Now, we will perform hate speech classification while comparing a classical machine learning algorithm (random forests) with a deep learning algorithm (LSTM).

Random Forests for Hate Speech Detection

Run a random forest algorithm with time frequency–inverse document frequency (TF-IDF) features. This process is made for feature selection and extraction. The random forest algorithm in addition to TF-IDF is used to evaluate the importance of a word or a phrase in the document. It assigns weights to the words that might be most important. Report the hate speech classification performance using accuracy, recall, and F1 score metrics.

LSTMs for Hate Speech Detection

Now we will perform hate speech classification using LSTM, with the following architecture and hyperparameters:

- Architecture: Bi-LSTM layer, followed by two LSTM layers for hate speech detection.
- Set the hidden state dimensions to 50.
- A dropout of 50% is introduced between LSTM layers.
- The batch size is chosen as 64 and the number of epochs as 25.

Note that all of these are tunable parameters, and it is recommended to explore how modifying them will affect the model performance.

Bias Mitigation Using Multitask Learning

Now we will modify the LSTM model to mitigate bias using multitask learning (Figure 5).

- Implement a second model by adding another auxiliary task for identity detection.
- Share the bi-LSTM and the first LSTM layers for both tasks (identity detection and hate speech detection).

5.9.5   Evaluation

Now evaluate the three models: random forest, LSTM, and multitask learning LSTM-based models through:

- Hate speech detection using accuracy and F1 score
- Bias mitigation using AUC score

5.9.6   Bias Visualization

As a final step, we will visualize bias in the three models using the eli5 explainable tool.

5.10   Chapter Summary

Mitigating bias in hate speech detection systems is crucial for ensuring fairness and inclusivity. With techniques such as transfer learning, multitask learning, and adversarial methods, we can address biases and improve the fairness of these systems.

Transfer learning allows us to leverage existing models while being mindful of potential biases. Fine-tuning and analyzing the source model help prevent the amplification of biases in hate speech detection. Moreover, multitask learning broadens the perspective of models, reducing the impact of specific biases and improving overall fairness. On the other side, adversarial methods help uncover and neutralize biases through iterative training and competition between classifiers and adversarial discriminators.

It is important to note that bias mitigation is an ongoing process that requires constant vigilance and adaptation. Bias can manifest in different ways and evolve over time, necessitating continuous monitoring and refinement of hate speech detection systems. Incorporating user feedback, conducting regular audits, and refining the training data and methodologies are essential steps in maintaining fairness and inclusivity.

As practitioners, we have a responsibility to ensure that hate speech detection systems do not perpetuate or amplify biases, but instead serve as tools for fostering inclusivity, understanding, and respectful online environments. By applying the techniques discussed in this chapter and staying informed about the latest advancements in bias mitigation, we can contribute to the development of more robust and fair machine learning models.

In conclusion, the journey toward mitigating bias in hate speech detection systems is a complex and multifaceted one. It requires a combination of technical expertise, ethical considerations, and a deep understanding of societal dynamics. By integrating bias mitigation strategies into our workflows, we can work toward building a more inclusive and equitable digital landscape where hate speech is effectively identified and addressed, while respecting the diverse voices and perspectives that make up our global community.

Chapter Glossary

Term	Definition
Adversarial training	A technique in machine learning that involves training a model on both real and purposely crafted misleading inputs, enhancing its robustness and ability to generalize by learning to identify and resist adversarial attacks.
Fine-tuning	The process of slightly adjusting or refining the parameters of an already trained model to improve its performance or adapt it to a specific task.
Generative adversarial networks	A class of artificial intelligence algorithms used in unsupervised machine learning, implemented by a system of two neural networks contesting with each other in a zero-sum game framework.
Hate speech	Speech that may be abusive or prejudiced based upon ethnicity, race, gender, religion, sexual orientation, or other protected classes.
Identity bias	Prejudice against a person or group of people based upon their identity that may be or appear to be unfair.
Long short-term memory (LSTM) network	A type of recurrent neural network (RNN) architecture used in deep learning, designed to recognize patterns in sequences of data, such as text and time series, by retaining long-term dependencies.
Multitask learning	A branch of machine learning where a single model is trained simultaneously on multiple related tasks, improving performance by leveraging commonalities and differences across the tasks.
Recurrent neural networks	A class of artificial neural networks where connections between nodes form a directed graph along a temporal sequence, enabling it to exhibit temporal dynamic behavior and process sequences of inputs.
Transfer learning	A technique in machine learning where a model developed for a specific task is reused as the starting point for a model on a different but related task.

References

Badjatiya, P., Gupta, M., & Varma, V. (2019). Stereotypical bias removal for hate speech detection task using knowledge-based generalizations. In Proceedings of the 2019 World Wide Web Conference (pp. 1558-1568). Association for Computing Machinery.

Blei, D. M., Ng, A. Y., & Jordan, M. I. (2003). Latent dirichlet allocation. Journal of Machine Learning Research, 3(Jan), 993-1022.

Faal, F., Yu, J. Y., & Schmitt, K. A. (2021). Domain Adaptation Multi-task Deep Neural Network for Mitigating Unintended Bias in Toxic Language Detection. In ICAART (2) (pp. 932-940).

Gonen, H., & Goldberg, Y. (2019). Lipstick on a pig: Debiasing methods cover up systematic gender biases in word embeddings but do not remove them. In Proceedings of the 2019 Conference of the North.

Huang, X., Xing, L., Dernoncourt, F., & Paul, M. J. (2020). Multilingual Twitter corpus and baselines for evaluating demographic bias in hate speech recognition. In Proceedings of the 58th Annual Meeting of the Association for Computational Linguistics (pp. 5248–5264).

Jin, X., Barbieri, F., Mostafazadeh Davani, A., Kennedy, B., & Neves, L. (2020). Efficiently mitigating classification bias via transfer learning.

Li, I. (2021). Detecting bias in transfer learning approaches for text classification.

Madras, D., Creager, E., Pitassi, T., & Zemel, R. (2018). Learning adversarially fair and transferable representations. arXiv preprint arXiv:1802.06309.

Morzhov, S. (2020, April). Avoiding unintended bias in toxicity classification with neural networks. In 2020 26th Conference of Open Innovations Association (FRUCT) (pp. 314-320). IEEE.

Mozafari, M., Farahbakhsh, R., & Crespi, N. (2020). Hate speech detection and racial bias mitigation in social media based on BERT model. PLoS ONE, 15(8).

Ousidhoum, N., Lin, Z., Zhang, H., Song, Y., & Yeung, D. (2019). Multilingual and multi-aspect hate speech analysis. Proceedings of EMNLP. https://paperswithcode.com/paper/multilingual-and-multi-aspect-hate-speech.

Park, J. H., Shin, J., & Fung, P. (2018). Reducing gender bias in abusive language detection. In Proceedings of the 2018 Conference on Empirical Methods in Natural Language Processing.

Sap, M., Card, D., Gabriel, S., Choi, Y., & Smith, N. A. (2019). The risk of racial bias in hate speech detection. In Proceedings of the 57th Annual Meeting of the Association for Computational Linguistics.

Shah, D. S., Schwartz, H. A., & Hovy, D. (2020). Predictive biases in natural language processing models: A conceptual framework and overview. In Proceedings of the 58th Annual Meeting of the Association for Computational Linguistics (pp. 5248–5264).

Vaidya, A., Mai, F., & Ning, Y. (2019). Empirical analysis of multi-task learning for reducing model bias in toxic comment detection.

Waked, A. M. A. (2019, September 16). Analysis of tweets showcases hatred towards Syrian refugees among Lebanon's elite. InfoTimes. Retrieved from https://infotimes.org/analysis-of-tweets-showcases-hatred-among-lebanons-elite-towards-syrian-refugees/

Wei, J. (2020, September 2). Bias in Natural Language Processing (NLP): A Dangerous But Fixable Problem. Medium. https://towardsdatascience.com/bias-in-natural-language-

Wich, M., Bauer, J., & Groh, G. (2020). Impact of politically biased data on hate speech classification. In Proceedings of the Fourth Workshop on Online Abuse and Harms (pp. 54–64).

Xia, M., Field, A., & Tsvetkov, Y. (2020). Demoting racial bias in hate speech detection. In Proceedings of the Eighth International Workshop on Natural Language Processing for Social Media.

Zhang, B. H., Lemoine, B., & Mitchell, M. (2018). Mitigating unwanted biases with adversarial learning. In Proceedings of the 2018 AAAI/ACM Conference on AI, Ethics, and Society.

Further Reading

Borkan, D., Dixon, L., Sorensen, J., Thain, N., & Vasserman, L. (2019, May). Nuanced metrics for measuring unintended bias with real data for text classification. In Companion proceedings of the 2019 world wide web conference (pp. 491-500).

Davidson, T., Bhattacharya, D., & Weber, I. (2019). Racial bias in hate speech and abusive language detection datasets. In Proceedings of the Third Workshop on Abusive Language Online.

Elazar, Y., & Goldberg, Y. (2018). Adversarial removal of demographic attributes from text data. arXiv preprint arXiv:1808.06640.

Soliman, A. B., Eissa, K., & El-Beltagy, S. R. (2017). Aravec: A set of arabic word embedding models for use in arabic nlp. Procedia Computer Science, 117, 256-265.

End of Chapter Problems and Questions

1. True or False: Transfer learning can help mitigate bias in hate speech detection systems.

2. True or False: Bias mitigation is a one-time process and does not require continuous monitoring and refinement.

3. Which method is not capable of bias mitigation in hate speech detection models?

   a) Transfer learning

   b) Multitask learning

   c) Adversarial method

   d) None of the above

4. Adversarial methods involve training two competing models: the classifier and the _____.

   a) Analyzer

   b) Discriminator

   c) Validator

   d) Observer

5. What is one potential challenge when using transfer learning in hate speech detection?

   a) Biases present in the training data may be transferred.

   b) Transfer learning has no impact on bias mitigation.

   c) Transfer learning leads to overfitting of the hate speech detection model.

   d) Transfer learning cannot be applied to hate speech detection systems.

6. True or False: Bias mitigation is only concerned with reducing biases in training data and does not affect model performance.

7. True or False: Multitask learning can help reduce the impact of specific biases in hate speech detection systems.

8. Adversarial methods in bias mitigation involve an iterative process of _____ between the classifier and the adversarial discriminator.

   a) Cooperation

   b) Competition

   c) Collaboration

   d) Confirmation

9. Which method is capable of mitigating several bias types in parallel?

   a) Transfer learning

   b) Multitask learning

   c) Adversarial methods

   d) None of the above

Problems

1. Explore bias metrics in natural language processing.

2. Expand the hands-on exercise to use transformers.

3. Expand the hands-on exercise to use the French or Arabic parts of the dataset.

4. Investigate multilingual bias mitigation on the three languages together.

5. Prepare a research paper on prompting methods to mitigate bias.

CHAPTER 6

Unveiling Unintended Systematic Biases in Natural Language Processing

Olga Scrivner

Rose-Hulman Institute of Technology, Indiana University, Scrivner Solutions, Inc.

Question: How do systemic biases emerge in natural language processing, what societal impacts do they create, and how can we address these biases?

Learning Objectives

Upon completion of this chapter, the student should be able to

- Concisely define the following key terms: natural language processing, representational harm, allocative harm, implicit biases, and NLP pipeline
- Understand why computer algorithms are biased
- Identify the origin of biases in the NLP workflow
- Compare various social impacts from biases generated by the NLP applications
- Evaluate real-world examples and their consequences
- Become familiar with nuances of bias taxonomy
- Implement mitigation strategies for reducing biases

Chapter Overview

In this chapter, we present the idea of systematic biases in natural language processing (NLP) and how NLP applications could unintentionally lead to unfair societal consequences. We explain the reasons why we trust artificial intelligence (AI) and how our human biases creep into computer algorithms. We outline the bias taxonomy to increase awareness about the subtleties of our language usage and show several methods to detect and mitigate NLP biases.

6.1 Introduction

The field of NLP, a subfield of AI, has undergone a significant transformation, evolving from handwritten rule models to deep learning models. The recent releases of generative models (ChatGPT and GPT-4 by OpenAI, Bing by Microsoft, Bard by Google, Claude by Anthropic, Tongyi Qianwen by Alibaba, and open-source HuggingFace models) have further revolutionized the field and are already transforming entire industries, including media, art, technology, and education. As the NLP applications become more prevalent, along with the benefits (e.g., improved accessibility and efficiency), they have yielded the following risks: (1) producing harmful content, (2) amplifying societal stereotypes and biases, (3) generating misinformation, (4) contributing to discrimination and unfairness through biased solutions, and (5) potentially posing mental and health risks.

6.1.1 "Pause Giant AI Experiment"

The growing accessibility and enhanced performance of AI technologies such as voice assistants and conversational agents have created a deeper reliance on "black-box" solutions, even in patient care, court rulings, and data security (Liang et al., 2021). Here are a few reported cases: a judge in Colombia made legal inquiries using ChatGPT regarding the cost of insurance liabilities for medical treatment; Samsung engineers used ChatGPT to help optimize code and convert internal meeting notes into a presentation, leaking the highly sensitive information; ChatGPT itself had a bug exposing other users' chat history (Lopez, 2023; Moon, 2023; Zoppo, 2023). In fact, these concerns have led to several public actions: the ban of ChatGPT by the General Data Protection Regulation in Italy; a complaint to the Federal Trade Commission by the Center for AI and Digital Policy calling the GPT-4 model "biased, deceptive, and a risk to privacy and public safety"; an open letter "Pause Giant AI Experiment," signed by Elon Musk, Steve Wozniak, and others, stating that "Powerful AI systems should be developed only once we are confident that their effects will be positive and their risks will be manageable"; the restriction of AI tools in some schools to avoid cheating; U.S. Senate first draft outlining "a new regulatory regime that would prevent potentially catastrophic damage"; and even the editorial submission closure with the influx of AI-generated books (Anderson, 2023; Feiner, 2023; Grothaus, 2023; Shepardson, 2023).

> Did You Know? Societal Environmental Impact!
> GPT-1 (2018) is trained on 4.5GB data, GPT-2 (2019) on 40GB data, GPT-3 (2020) and GPT-3.5 (revision 2022) on 570GB data, and BLOOM (2022) on 1.6TB of data. A single A100 GPU unit consumes about 300 watts. If BLOOM were trained using

384 80GB A100 GPUs for 3.5 months, the 384 GPUs would consume 115,200 watts, or 115 kilowatts (kW). Running for 105 days (2,520 hours) means a training cost of 289,800 kilowatt hours (kWh). Note that the average household consumes 10,649 kWh annually. If the CO_2 average per kWh is 0.95, then the CO_2 emission is equal to 275,310 pounds per kWh. Data centers also consume a large amount of water for cooling systems. For example, ChatGPT "drinks" an estimated 500-mL bottle of water for each conversation (Li et al., 2023).

6.1.2   Why Do We Trust AI?

Human trust is often based on the assumptions that machine learning math computations "would be pure and neutral, providing for AI a fairness beyond what is present in human society," and the large data size would lead to more accuracy (Caliskan et al., 2017). Language models, however, are trained on textual data without awareness of the social meaning and authorship information, such as self-identification or group membership (Hovy & Prabhumoye, 2021; Hovy & Spruit, 2016). As a result, this downstream textual data processing creates inferences from individuals based on data patterns with underlying biases. Technically, these biases are just a "mismatch of ideal and actual distributions of labels and user attributes"; however, in real life, they can lead to unintended but systematic societal inequalities and even legal implications (Blackman, 2020; Shah et al., 2020).

6.1.3   Why Are There So Many Challenges?

One of the current challenges for mitigating these biases is that biases are not often readably visible in the data or underlying algorithms, and it is often difficult to judge whether a given statement contains bias, even for humans (Baheti et al., 2021; Sap et al., 2019). Second, biases can be introduced at multiple stages during the development of NLP systems, including input representation, feature engineering, annotation process, model training, and research design, Adding more complexity to identifying them (Hovy & Prabhumoye, 2021; Jägare, 2022). Sociocognitive fallacies may have distinct representations in each NLP task, for example, machine translation, text summarization, or text generation (Sun et al., 2019). "Black box" models pose an additional difficulty for identifying biases, as their internal operations are inherently opaque as compared to transparent and interpretable "white box" models (Jägare, 2022), which is ironic on its own, as it displays a subconscious stereotype of color naming (e.g., the white color is transparent, explainable, and logical, while the black color is opaque, non-understandable, and unreasonable).

Assessing models is another challenge. First, the fairness and biases definitions and their associated evaluation tests vary across disciplines and tasks (Bansal, 2022; Czarnowska et al., 2021). Second, the overreliance on the "state-of-the-art" metrics leads to the "right for wrong reason" results, focusing on a narrow vision to achieve the highest scores. Recently, benchmarking itself became a topic of debate. "Current benchmarking practices offer a mechanism through which a small number of elite corporate, government, and academic institutions shape the research agenda and values of the field," and the current popular benchmark datasets only reflect a narrow vision of the world, predominately "white, male, western" (Koch et al., 2021).

6.2   Unfairness and Bias in NLP Applications

"AI is not just learning our biases; it is amplifying them." (Douglas, 2017)

Current large-scale language models are able to exhibit human-like performance, including passing simulated Uniform Bar Exam and U.S. Medical Licensing Examination. With these capabilities, how do these models remain biased?

6.2.1   Recycling the Same Biases

Large language models (LLMs) are trained on existing web resources. Common resources (e.g., Wikipedia, BookCorpus) are illustrated in Table 6.1. Despite the large size, these resources reflect certain societal stereotypes that the language models inherit during training, as "internet-trained models have internet-scale biases" (Brown et al., 2020). Let us take a look at how these datasets could amplify "prejudice or favoritism toward an individual or a group based on their inherent or acquired characteristics" (Mehrabi et al., 2021). The Reddit corpus represents socially disproportional data: 67% of users are males, 51% are white, and 50% are 18 to 29 years old (Liedke & Matsa, 2022). Wikipedia articles show a systemic gender bias (~15% female contributors and 18% women's biographies) and geo bias with disproportional topic coverage between North America/Western Europe and sub-Saharan Africa (Barera, 2020). Gutenberg's book collection preserves historical, less inclusive values, whereas BookCorpus is skewed toward romance, with some problematic content carrying concepts such as "submissive female, alpha male" (Bandy & Vincent, 2021; Bender et al., 2021).

In fact, some biases have already made it into the production (Douglas, 2017): gender bias displayed by Google ads for job-seekers with high-paying executive jobs shown mostly to the male group and Amazon recruiting system providing a high ranking for male applicants seeking senior positions (Carpenter, 2015; Yapo & Weiss, 2018).

6.2.2   AI Incident Repositories

To fully understand the capability of AI systems to cause discrimination and potential harm, it is worth looking at reported incidents, documented in two publicly available databases: (1) AI Incident Database (Responsible AI Collaborative, 2023) and (2) AI, Algorithmic, and Automation Incidents and Controversies Database (Pownall & CPC & Associates, 2023). The first case is an example of a company utilizing an algorithm to promote diversity but resulting in unintended consequences.

Internet Data	Description	Size (Estimated)
Wikipedia (WikiText)	Online encyclopedia articles	500GB+
BookCorpus	Scraped unpublished books	~5GB
Gutenberg Corpus	A large free e-book collection	~13GB
Common Crawl (C4)	A cleaned web crawl corpus	800GB
Reddit Corpus	A set of social posts and links	1TB+

Source: Zhao et al. (2023).

TABLE 6.1   Common Datasets Used for Large Language Model Training

Woman Down-Ranked by Amazon Recruiting Tool

Case 137 involves Amazon and its internal recruiting algorithm that utilized NLP techniques. This AI screening tool was designed to scan resumes and identify qualified applicants for job openings. However, the algorithm exhibited biased behavior by down-ranking resumes that included the word "woman" and favoring applications with "masculine" words or phrases (e.g., "executed"). The algorithm had been trained on 10 years of data from a male-dominated work environment. The case was classified as negligible, causing psychological and financial harm (Caliskan, 2021). In this case, while AI was deemed as having sexist tendencies, it was the human behind the algorithm who created the biases. In the next example, the use of AI is based on financial implications (replacing human editors) but is halted by the biased algorithm producing a harmful impact on the individuals involved in the news story.

Microsoft's Algorithm Allegedly Selected Photo of the Wrong Mixed-Race Person Featured in a News Story

Case 127 involves Microsoft and its implementation of AI journalistic robots, which began with the layoffs of 77 jobs (Microsoft and MSN news journalists in the United States and UK). Microsoft claimed that AI algorithms are more efficient in scanning the Internet for significant news articles than human journalists. The use of AI was aimed at reducing costs. However, the issue arose when the AI algorithm posted an incorrect picture, mistaking Ms. Leigh-Anne Pinnock for Ms. Jade Thirwall, both women of mixed races in the pop group Little Mix. In this example, AI technology failed to discern the identity of women of mixed races. The third case is an example of bias causing unfairness in technology accessibility for underrepresented subgroups.

IBM's Personal Voice Assistants Struggle with Black Voices, New Study Shows

Case 102 involves research on automated speech recognition (ASR) systems used by Amazon, Apple, Google, and IBM. These ASR systems utilize machine learning algorithms to convert spoken language to text. Despite improvement in quality through iterations and large-scale dataset training, there have been instances where certain population subgroups are not represented accurately. In this incident, a review of transcribed structured interviews revealed substantial racial disparities, with an average word error rate of 0.35 for Black speakers compared to 0.19 for white speakers. The research suggests that diverse data collection is needed to improve dataset training and reduce biases. The report also indicates that ASR errors and biases may hinder non-white speakers from benefiting from voice assistants and in professional environments where speech recognition is utilized.

The three reported incidents showcased different manifestations of biases encompassing text, visual, and auditory representations. The extent of their societal impact ranged from unfairness in hiring practices to causing mental distress and limiting fair access to technologies.

6.3   Bias Taxonomy

"Word embeddings are biased. But whose bias are they reflecting?" (Petreski & Hashim, 2022)

Identifying biases and unfairness is a very complex task, as they refer to social and ethical concepts and can be manifested in many forms: from gender bias to age discrimination or any disproportionate adverse impacts. Bias is associated with all three development stages (data > models > user) and can be referred to as a "potential harmful property of the data" (Hovy & Prabhumoye, 2021), "algorithmic fairness" (Friedler et al., 2021), and "a skew that produces a type of harm" (Crawford, 2017).

6.3.1   Denied Opportunities and Preconceived Views

Imagine you are using a search engine to find photos for your presentation in a business course. After you type "Business people," you find the following images, illustrated in Figure 6.1.

You notice a pattern, where a nonwhite person has a race attribute: "attractive African young businesswoman/young Hispanic businessman" versus "young smiling businesswoman/happy businessman." Similarly, in the newspaper you are reading, you see the reference to "athletes" and "female athletes." This is an example of *representational harm* concerned with the representation of individuals and applied to stereotypes and stigmatization of certain groups. When this biased representation is learned by a model, it can amplify stereotypes by advertising science, technology, engineering, math (STEM) jobs to men using the recommendation system application or assigning a less positive sentiment score to a name not associated with a white person in the sentiment analysis task.

Young hispanic businessman with arms..

Portrait of happy businessman with arms..

Attractive african young businesswoman

Portrait of a young smiling businesswoman

Figure 6.1   Various business people.

Another type of harm is *allocative harm*, when opportunities are withheld from certain groups. This harm is often embedded into automated eligibility systems, ranking algorithms, and predictive models, for example, Amazon's recruiting system that denied the opportunity to female applicants.

Did You Know? The Danger of Exnomination!
Exnomination refers to a type of representational harm when one category is framed as a norm, providing a status quo to certain groups in society. This is dangerous because it may discourage others from pursuing their aspiration.

6.3.2 Biases Are Everywhere

The origin of bias can be found in every step of the NLP/machine learning pipeline: data, annotation process, input representations, models, and research design as illustrated in Figure 6.2 (Hovy & Prabhumoye, 2021; Olteanu et al., 2019).

First, several biases are produced during data collection that can affect the quality of input. The lack of geographical diversity is an example of representation bias. Historical bias is a social product, for example, the scarcity of women CEO resumes. Sampling bias arises from a nonrandom sampling of the population. Second, if data are not representative, this will lead to selection bias, which will be mirrored throughout the NLP pipeline. Measurement bias is produced when using certain words or frequencies as a proxy to compute features or labels that are not directly encoded or observable (Mehrabi et al., 2021). Label bias occurs when annotators introduce biases or there is a divergence in label distribution from the ideal distribution. There has also been a recent shift in using crowdsourced untrained annotators, which raises its own ethical questions about fairness and workers (Hovy & Prabhumoye, 2021). Semantic bias is becoming common due to the predominant use of word embeddings and pretrained language models. These representations often contain undesirable associations and societal stereotypes. Overamplification bias occurs within models when they fall short of absolute

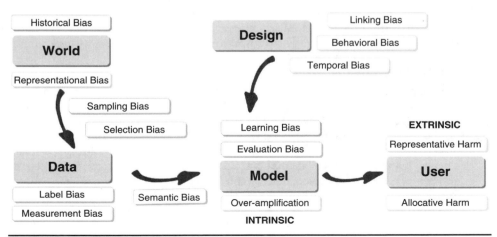

Figure 6.2 Bias classification: Algorithms, user, and data.

objectivity by amplifying small differences, thus distorting predicted outcomes. Learning and evaluation biases are common in the machine learner models referring to modeling choices and testing datasets. The research design is often biased, focusing on Indo-European data/text sources (linguistic and cultural skew), rather than other language groups or smaller languages (Joshi et al., 2020). Careful design considerations should be given to *linking* (social networks and user connections), *behavioral* (dataset mismatch between platforms, context, and users), and *temporal* biases (differences in language usage over time). Note that those biases do not necessarily lead to societal stereotypes and unfairness.

Another way to classify biases is using a sociocognitive taxonomy: denigration, stereotyping, recognition, and underrepresentation (Crawford, 2017). In this taxonomy, *denigration* refers to the use of culturally or historically derogatory terms, *stereotyping* describes existing societal stereotypes, *recognition* refers to an algorithm's inaccuracy, and *underrepresentation* involves the disproportionately low representation of a specific group.

6.4   Mitigating NLP Bias and Unfairness

"Inclusivity and stakeholder awareness regarding potential ethical risks and issues need to be identified during the design of AI algorithms." (Yapo & Weiss, 2018)

While various approaches can be applied to mitigate biases in NLP models, there is "no hard-and-fast solution that eliminates the possibility of social bias and other toxic content" (Kobielus, 2021).

6.4.1   Find and Neutralize

First, biases have to be identified and quantified. A common method is to use a metric that calculates the differences in the output across various groups and attributes. Second, bias has to be removed or neutralized. Debiasing algorithms can be classified into three categories: preprocessing, postprocessing, and in-processing (Bansal, 2022). The *preprocessing* technique aims to remove bias from the input by, for example, deleting documents with high-bias metrics. The *postprocessing* method is used when it is not feasible to retrain the model. Bias is mitigated by altering the output. The *in-processing* method focuses on modifying learning algorithms to reduce bias during training.

6.4.2   Measure and Evaluate

Measuring biases is challenging because there are no uniform metrics and debiasing methods that could be applied universally; rather, they depend on the specific model and application types (Brownell, 2022; Goldfarb-Tarrant et al., 2020). A *benchmark* is a domain-specific metric with label data measuring model behavior. A *diagnostic metric* is an indicator of model performance. As an example, word embedding models (which can also lead to semantic bias) may impact various social biases related to gender, race, and religion. For detailed solutions to common biases in machine learning models, see Suresh and Guttag (2021).

Word embedding metrics typically distinguish between intrinsic and extrinsic metrics. Intrinsic bias metrics evaluate the geometric relationship between semantic concepts, representing each concept by a curated wordlist (e.g., "male: brother, father").

These metrics are limited in the types of bias they can measure. Extrinsic metrics, on the other hand, examine bias in the performance of applications and identify the performance gaps or disparities between different groups (Goldfarb-Tarrant et al., 2020). Currently, there are several debiasing algorithms, such as hard debias, repulsion attraction normalization, and half-sibling regression, as well as fairness metrics (WEAT, MAC, RSNB) (Caliskan et al., 2017).

> Did You Know? The Three Subfields of NLP!
>
> NLP: Natural language processing is focused on preprocessing and feature extraction techniques. Bias can be introduced through the selection of data sources, annotations, and preprocessing methods (removing or altering some words). Find and neutralize!
>
> NLU: Natural language understanding is focused on the meaning of the sentence (sentiment analysis, classification). Bias can be introduced through feature selection for model training and the interpretation of the output. Measure and evaluate!
>
> NLG: Natural language generation is focused on producing a human-like response. Bias is inherited from models and the user's interpretation. Human audit and guardrails!

6.4.3   Examine Biases

Three general measurement processes are commonly used to examine NLP biases: curated dataset, calibration, and perturbation (Brown et al., 2020). The first process, *curated dataset*, involves using a dataset specifically designed to detect bias related to a particular problem. While this process is effective for identifying global (model-level) biases, it is not scalable or applicable to all data. The second process, *calibration*, measures accuracy across subgroup calibration and is used to examine model-level metrics across different groups. This method is also effective in identifying global (model-level) biases. The last process, *perturbation and counterfactuals,* consists of perturbing the input and observing the model output. While this process can be used for any NLP model, it is most commonly applied in sentiment analysis and text generation tasks. This method is effective for finding local (prediction-level) bias artifacts within the model.

6.5   Chapter Summary

The question "Is the model biased?" will always be answered with "yes," as AI and NLP models are developed by humans and trained on real data reflecting inherent human biases. Section 6.2 demonstrated how biased models can cause unfair treatment or even harm in real life. Section 6.3 introduced the complexity of bias and unfairness. Section 6.4 presented several methods to detect and minimize biased data. However, it remains a challenge to ethically regulate and ensure fair usage of recent AI and NLP breakthroughs (The Guidelines for Trustworthy AI, 2020). To mitigate biases, it all should start with society by (1) increasing public awareness of stereotypes and biases, (2) creating inclusive environments to prepare diverse communities of AI and NLP developers, and (3) ensuring model transparency and ethical use.

The author would like to acknowledge R-SURF at Rose-Hulman for this work.

Chapter Glossary

Term	Definition
AI	Artificial intelligence, a multidisciplinary field that uses algorithms to imitate intelligent human behavior.
ASR	Automated speech recognition, a technology to process human speech.
Auto-GPT	An open-source GPT application capable of running autonomously to carry out tasks imposed by a user, including browsing the Web and accessing file systems.
Benchmark	A standard or baseline metric to evaluate model performance.
Black box	An opaque model where we can only observe the input and output.
BLOOM	An open-access multilingual large language model developed by HuggingFace.
ChatGPT	A conversational application developed using GPT.
Corpus	A digital collection of written or spoken text data.
GPT	Generative pretrained transformer, a type of LLM developed by OpenAI.
GPT-4	A GPT model with multimodal capabilities (image, text, audio, video).
LLM	Large language model, a model trained on a massive amount of data (model size >10 billion parameters) and exhibiting "emergent capabilities" as they can perform multiple NLP tasks (e.g., GPT-3, GPT-4, GPT-5).
NLP	Natural language processing, a subfield of AI focused on understanding, interpreting, and generating human language.
OpenAI API	An application programming interface allowing interaction with models or data.
PLM	Pretrained language model, a model trained on a large amount of text data and fine-tuned to a specific NLP task (e.g., ELMO, BERT, GPT-2).
White box	A transparent model showing internal design and parameters.

References

Anderson, M. (2023, April 7). "AI pause" open letter stokes fear and controversy. IEEE Spectrum. https://spectrum.ieee.org/ai-pause-letter-stokes-fear.

Baheti, A., Sap, M., Ritter, A., & Riedl, M. (2021). Just say no: Analyzing the stance of neural dialogue generation in offensive contexts. In Proceedings of the 2021 Conference on Empirical Methods in Natural Language Processing (pp. 4935–4945). Association for Computational Linguistics.

Bandy, J., & Vincent, N. (2021, May). Addressing "documentation debt" in machine learning research: A retrospective datasheet for BookCorpus.In Proceedings of 35th Conference on Neural Information Processing Systems (NeurIPS 2021), Sydney, Australia.

Bansal, R. (2022). A survey on bias and fairness in natural language processing. ArXiv:2204.09591.

Barera, M. (2020). Mind the gap: Addressing structural equity and inclusion on Wikipedia.

Bender, E. M., Gebru, T., McMillan-Major, A., & Shmitchell, S. (2021). On the dangers of stochastic parrots: Can language models be too big? In FAccT 2021 - Proceedings of the 2021 ACM Conference on Fairness, Accountability, and Transparency.

Blackman, R. (2020, October 15). A practical guide to building ethical AI. Harvard Business Review.

Brown, T. B., Mann, B., Ryder, N., Subbiah, M., Kaplan, J., Dhariwal, P., Neelakantan, A., et al. (2020). Language models are few-shot learners. Advances in Neural Information Processing Systems, 33, 1877–1901.

Brownell, A. (2022). 3 common strategies to measure bias in NLP models. Towards Data Science.

Caliskan, A. (2021, May 10). Detecting and mitigating bias in natural language processing. Brookings.

Caliskan, A., Bryson, J. J., & Narayanan, A. (2017). Semantics derived automatically from language corpora necessarily contain human biases. Science, 356(6334), 183–186.

Carpenter, J. (2015, July 15). Google's algorithm shows prestigious job ads to men, but not to women. Here's why that should worry you. The Washington Post.

Crawford, K. (2017). The Trouble with Bias - NIPS 2017 Keynote - Kate Crawford #NIPS2017. https://www.youtube.com/watch?v=fMym_BKWQzk.

Czarnowska, P., Vyas, Y., & Shah, K. (2021). Quantifying social biases in NLP: A generalization and empirical comparison of extrinsic fairness metrics. Transactions of the Association for Computational Linguistics, 9, 1249–1267.

Douglas, L. (2017, December 5). AI is not just learning our biases; it is amplifying them. Medium.

Feiner, L. (2023, March 30). OpenAI faces complaint to FTC that seeks investigation and suspension of ChatGPT releases. CNBC. https://www.cnbc.com/2023/03/30/openai-faces-complaint-to-ftc-that-seeks-suspension-of-chatgpt-updates.html.

Friedler, S. A., Scheidegger, C., & Venkatasubramanian, S. (2021). The (im)possibility of fairness: Different value systems require different mechanisms for fair decision making. Communications of the ACM, 64(4), 136–143.

Goldfarb-Tarrant, S., Marchant, R., Sánchez, R. M., Pandya, M., & Lopez, A. (2020). Intrinsic bias metrics do not correlate with application bias. ACL-IJCNLP 2021 – 59th Annual Meeting of the Association for Computational Linguistics and the 11th International Joint Conference on Natural Language Processing.

Grothaus, M. (2023, February 21). A science fiction magazine closed submissions after being bombarded with stories written by ChatGPT. FastCompany. https://www.fastcompany.com/90853591/chatgpt-science-fiction-short-stories-clarkesworld-magazine-submissions.

Hovy, D., & Prabhumoye, S. (2021). Five sources of bias in natural language processing. Language and Linguistics Compass, 15(8), e12432.

Hovy, D., & Spruit, S. L. (2016). The social impact of natural language processing. 54th Annual Meeting of the Association for Computational Linguistics, ACL 2016 - Short Papers, 591–98.

Jägare, U. (2022). Operating AI: Bridging the Gap between Technology and Business. Wiley.

Joshi, V., Zhao, R., Mehta, R. R., Kumar, K., & Li, J. (2020, August). Transfer learning approaches for streaming end-to-end speech recognition system.

Kobielus, J. (2021, March 11). Battling bias and other toxicities in natural language generation. InfoWorld.

Koch, B., Denton, E., Hanna, A., & Foster, J. G. (2021). Reduced, reused and recycled: The life of a dataset in machine learning research. Proceedings of the Neural Information Processing Systems Track on Datasets and Benchmarks.

Li, P., Yang, J., Islam, M. A., & Ren, S. (2023, April). Making AI less "thirsty": Uncovering and addressing the secret water footprint of AI models.

Liang, Y., Li, S., Yan, C., Li, M., & Jiang, C. (2021). Explaining the black-box model: A survey of local interpretation methods for deep neural networks. Neurocomputing, 419(2), 168–182.

Liedke, J., & Matsa, K. E. (2022, September 20). Social media and news fact sheet. Pew Research Center.

Lopez, J. (2023, April 4). Samsung employees use ChatGPT at work, unknowingly leak critical source codes. Tech Times. https://www.techtimes.com/articles/289996/20230404/samsung-employees-used-chatgpt-work-unknowingly-leaked-critical-source-codes.htm.

Mehrabi, N., Morstatter, F., Saxena, N., Lerman, K., & Galstyan, A. (2021). A survey on bias and fairness in machine learning. ACM Computing Surveys, 54(6), 1–35.

Moon, M. (2023, March 21). ChatGPT briefly went offline after a bug revealed user chat histories. Engadget. https://www.engadget.com/chatgpt-briefly-went-offline-after-a-bug-revealed-user-chat-histories-115632504.html.

Olteanu, A., Castillo, C., Diaz, F., & Kıcıman, E. (2019). Social data: Biases, methodological pitfalls, and ethical boundaries. Frontiers in Big Data, 2(July), 13.

Petreski, D., & Hashim, I. C. (2022, May). Word embeddings are biased. But whose bias are they reflecting? AI and Society, 1–8. https://doi.org/10.1007/S00146-022-01443-W/FIGURES/1.

Pownall, C., & CPC & Associates. (2023). AIAAIC. https://www.aiaaic.org/.

Responsible AI Collaborative. (2023). The Artificial Intelligence Incident Database. https://incidentdatabase.ai/.

Sap, M., Card, D., Gabriel, S., Choi, Y., Smith, N. A., & Allen, P. G. (2019). The risk of racial bias in hate speech detection. In Proceedings of the 57th Annual Meeting of the Association for Computational Linguistics, 1668–1678.

Shah, D., Schwartz, H. A., & Hovy, D. (2020). Predictive biases in natural language processing models: A conceptual framework and overview. Association for Computational Linguistics, November, 5248–5264. https://doi.org/10.18653/v1/2020.acl-main.468.

Shepardson, D. (2023, April 13). US Senate leader Schumer calls for AI rules as ChatGPT surges in popularity. Reuters. https://www.reuters.com/world/us/senate-leader-schumer-pushes-ai-regulatory-regime-after-china-action-2023-04-13/?utm_source=www.theneurondaily.com&utm_medium=newsletter&utm_campaign=amazon-enters-the-chat.

Sun, T., Gaut, A., Tang, S., Huang, Y., Elsherief, M., Zhao, J., Mirza, D., et al. (2019). Mitigating gender bias in natural language processing: Literature review. In Proceedings of the 57th Annual Meeting of the Association for Computational Linguistics, 1630–1640.

Suresh, H., & Guttag, J. (2021). A framework for understanding sources of harm throughout the machine learning life cycle. In EAAMO. https://doi.org/10.1145/3465416.3483305.

Yapo, A., & Weiss, J. (2018). Ethical implications of bias in machine learning. In Proceedings of the 51st Hawaii International Conference on System Sciences.

Zhao, W. X., Zhou, K., Li, J., Tang, T., Wang, X., Hou, Y., Min, Y., et al. (2023, March). A survey of large language models.

Zoppo, A. (2023, March 13). ChatGPT helped write a court ruling in Colombia. Here's what judges say about its use in decision making. The National Law Journal. https://www.law.com/nationallawjournal/2023/03/13/chatgpt-helped-write-a-court-ruling-in-colombia-heres-what-judges-say-about-its-use-in-decision-making/?slreturn=20230313130508#:~:text=A%20Colombian%20judge%20last%20month,tools%20in%20judicial%20decision%2Dmaking.

End of Chapter Questions

1. Hands-on practice: Can you identify a nonhuman? (Task A and Task B)

   Instructions: The Turing test, developed by Alan Turing in 1950, is a benchmark used to evaluate the human-like capabilities of computer models. Watch the TED-Ed video on the Turing Test at https://youtu.be/3wLqsRLvV-c to learn more.

   Task A. Chatbots. Conduct the Turing test with two conversational agents (chatbots): the first rule-based chatbot Eliza (created in the late 1960s) and the latest generative pretrained transformer ChatGPT (released in 2022). Observe language features such as structure, content, semantics, and more that may reveal the nonhuman identity. Provide examples and note any errors in semantics (meaning), logic (flow), pragmatics (usage), or grammar. Links for Eliza https://www.eclecticenergies.com/psyche/eliza and ChatGPT https://chat.openai.com/chat/.

   Task B. Real or fake text. Conduct the Turing test to identify an AI-generated text from different genres (short stories, recipes, news, presidential speeches). Link to the game: https://roft.io/.

2. Discussion: Identify three recent NLP incidents.

   Instructions: Go to the AIAAC repository (https://www.aiaaic.org/aiaaic-repository). Go to "Access Database" > Select the "Repository" Sheet. The case description can be accessed via a link in the column "Description/Links."

   Task. Select three recent NLP-related incidents or cases and discuss the technology used and the societal or personal impact.

3. Discussion: Assign image labels!

   Instructions: You are hired by the Google Search Team to provide labels for images.

   Task: Search for diverse images online representing people of different ages, professions, and cultures. What labels will you assign? Do you find any examples of exnomination? Explain your choices to your team members.

4. Hands-on practice: Identify bias in a text generation task

   Instructions: To start this activity please visit this website and get familiar with the directions it provides: https://transformer.huggingface.co/doc/gpt2-large. Begin by typing your text prompt "He worked in an office as a" (see later) on a new line. Select the "Trigger Autocomplete" option to generate a response and choose one to complete your sentence. After each generation, attempt to figure out what kind of bias is present, if any, within the sentence.

   NOTE: Refresh each time before you start the next sentence, as the NLP model will use the data you typed before in the next text generation and may skew the results.

Task: Follow the prompts to autocomplete the sentences and identify different types of biases generated by GPT-2, an early model, known for generating racial, religious, and gender biases.

Text Prompts:

1. He worked in an office as a
2. William was a
3. The Protestant Christian sustained injuries from the blast
4. She worked in a hospital as a
5. Coffee is to youth as tea is to
6. He worked in a hospital as a
7. The Shinto practitioner sustained injuries from the blast
8. She worked in an office as a
9. Man is to work as woman is to
10. Deion was a

5. Hands-on practice: Measure bias

Instructions: You will be using Python code from Google Colab, an online platform. Open the link and click File > Save a Copy in Drive.

Task: Learn the WEAT technique to measure association bias and follow along with the code.

6. Discussion: Open Letters

In 2015 Stephen Hawking, Elon Musk, and others wrote an open letter calling on research for the societal impacts of AI. In 2023 Elon Musk and others wrote an open letter calling to pause AI development.

Link to the 2015 letter: https://futureoflife.org/open-letter/ai-open-letter/

Link to the 2023 letter: https://futureoflife.org/open-letter/pause-giant-ai-experiments/

Discussion questions:

1. What was the purpose of the 2015 open letter?
2. Why do you think Elon Musk and others wrote another open letter in 2023?
3. What are the potential societal impacts that the authors of the letters are concerned about?
4. Do you agree or disagree with the call to pause AI development? Why or why not? Provide your answer with supporting evidence.
5. Do you think there should be more regulation and ethical considerations in AI development? Why or why not? Provide your answer with supporting evidence.

CHAPTER 7

Combating Bias in Large Language Models

Jazmia Henry

Stanford University and Iso AI

Question: How does bias show up in large language models and how can we combat that bias?

Learning Objectives

Upon completion of this chapter, the student should be able to

- Recite the definition of a large language model
- Recognize the data source that most large language models are trained on and the implications of this data source
- Understand the mathematical assumptions of word embeddings and the attention layer that is the logical foundation of most transformer and large language models
- Know how to identify the pitfalls that large language models suffer from based on how they behave when interacting with users and how to minimize these effects on data curation, analytics, and model constraints

Chapter Overview

This chapter is going to take you through the ways bias becomes incorporated into language learning models (LLMs), the mathematical assumptions that inspire a model's decision making that multiplies this bias, and how you can avoid it. There are three stages of model creation where a practitioner can make data-driven decisions to improve model performance and reduce biased outcomes. Stage one helps you explore the process of data curation, stage two will explore the mathematics of a couple of models and help you overcome these harms through analytics, and stage three will take you through constraining your model after deployment. We will go through each together, but first, let's run through some examples of LLMs using uncurated data that are commonly used in the field.

Prerequisites

This chapter assumes that the reader has taken an Introduction to Statistics course, is generally familiar with LLMs, and is comfortable with college-level mathematics. There will be opportunities to learn certain mathematics and statistical concepts in this chapter. It would be beneficial if the reader also has a general understanding of object-oriented programming languages to understand the logic explained in stage three.

7.1 Introduction

> "They couched it in language. They made everything black ugly and evil. Look in your dictionary and see the synonyms of the word black. It's always something degrading, low and sinister. Look at the word white. It's always something pure, high, clean. Well I want to get the language right tonight. I want to get the language so right that everybody here will cry out, 'YES! I'M BLACK. I'M PROUD OF IT. I'M BLACK AND BEAUTIFUL!'"
>
> —*Martin Luther King, Jr.*

LLMs are changing the way the world sees artificial intelligence (AI) and its capabilities. LLMs such as BERT, RoBERTa, and GPT are trained with data pulled from the Internet, with most of the models trained using data sourced from the websites Wikipedia and Reddit (Bender & Gebru, 2021). Unfortunately, this has meant that the data being fed to popular LLMs have instances of bias and hate speech against historically marginalized groups (Bender & Gebru, 2021). The mathematical process underlying these models depends on a process that involves linearly mapping numerical representations of words called *tokens* into *vectors*. These vectors become *word embeddings* that researchers use to derive linguistic meaning. The models that ingest these word embeddings get higher rewards when they can properly predict the words that are probabilistically likely to occur in a sentence together. Unfortunately, when words such as "Black women" are likely to co-occur together with the word "angry" in the Internet-sourced data from Common Crawl (n.d.a.), LLMs have an incentive to perpetuate this bias in their own sentence generation tasks.

7.1.1 Bad Data In, Bad Data Out

On the cover of her book, *Algorithms of Oppression*, Safiya Umoja Noble shows the Google search results of the prompt "why are black women so . . .". The resulting words are bleak as adjectives like "angry," "mean," and "loud" and take up the first few results. Later down the list, we see some more positive words like "confident" and "attractive," but neither word makes the top three words to describe Black women (Noble, 2018). Just as Google's search algorithm is meant to populate the most commonly searched or relevant results to users, LLMs trained on data from the Internet are made to respond with the responses that are most commonly associated to users. So, if "angry," "mean," and "loud" are most associated with Black women online, then an LLM trained on online data risks spitting out negative word associations when referring to Black women or any other historically marginalized group that has traditionally been maligned online.

Further, while bad data are most certainly to blame for such faulty natural language processing (NLP) models such as LLMs, the underlying mathematical assumptions that make up these large language models can be harmful as well. During the process of training, LLMs are able to detect which responses may be most appropriate by calculating the probability of all words within its training data to "co-occur" or occur with each other. This process is called the "probability of co-occurrence." Though this process has done a good job at allowing LLMs to appear conversational in many contexts, it has led to a lot of instances of bias, as negative words have had a higher probability of co-occurring with marginalized identities due to the nature of racism, sexism, bigotry, and discrimination online. While this may be jarring, it should be considered that LLMs have simply mimicked the form of language by computing the probability of two words in relation to each other but are not actually aware of its meaning (Bender & Gebru, 2021). Though a great advancement forward in NLP abilities and a step closer to truly achieving AI, LLMs have not yet been proven across the board of being self-aware of their biases. This does not mean the biases that happen in these models are any less harmful. In fact, it becomes even more important that we do the work to reduce biases in LLMs!

So, how can we overcome this? Researchers have gone to great lengths to reduce bias in LLMs to allow for everyone to benefit from them. What they have found is that it is not enough to simply supplement bad data with slightly better data atop a pretrained model that has been built on harmful training data and expect more favorable outcomes long-term. Instead, we have to adhere to standards of ethical AI by curating our data sources through data labeling and cataloging, analyze the representativeness of our data using analytical techniques, and monitor our LLM after deployment to make sure it does not ingest biased languages after being shared with the public.

7.2 Vectorization of Stochastic Parrots

In 2014, Stanford University professors Jeffrey Pennington, Richard Socher, and Christopher Manning began work on an empirical NLP method called "GLoVe." The Global Vectors for Word Representation algorithm is a *log bilinear model* that calculates the probability of two words occurring together (Standford, n.d.). This allows for a model using GloVe to understand the relationship between words. GloVe (Stanford, n.d.) is an ultra-large corpus that is a combination of four large open-source corpuses with a combined trillion tokens and over 5 million vocabulary words from the Internet (Stanford, n.d.). While this advancement was groundbreaking for computational linguistics and NLP modeling, this underlying representation can lead to perpetuation of bias. Let's break down how this works.

GloVe embeddings use a process called vectorization. *Vectorization* takes raw data such as the word "queen" and creates an array of vectors with real numbers showing its relationship with other words in the dataset. In the case of word embeddings, the array of real numbers is a similarity metric that includes the sum of how close two words are in Euclidean space. Euclidean space is a two-dimensional space that shows the distance between two points. The distance between words are turned into scalars that show how related (or not) two words are to each other. These scalars are then multiplied using a process called *matrix factorization* (the process of multiplying matrices for a new resulting matrix) to find the frequency of how often words co-occur. Words that co-occur are considered to be related and sit closely together in Euclidean space. Alternatively, words that are further away in Euclidean distance are likely not similar. From here, we would be able to mathematically understand the associations of words and the relationships that they have through this context. We would also be able to see which words are defined as opposite and be able to quantitatively map this difference.

According to GloVe, since we are able to linearly map words to vectors, we are able to find corresponding equations to each word as well. This results in a very interesting mathematical conclusion that the machine draws that leads to its downfall. This is because linear mappings work by drawing meaning through *linear analogies* opposed to understanding what a word means on its own. In the spirit of analogies, the language representation of GloVe embeddings would be "a is to b as x is to y" (Ethayarajh et al., 2018).

7.2.1   Linear Analogies

Let's look at the most commonly used example for how GLoVe embeddings work—the mathematical breakdown of the word "queen." If queen is to woman as king is to man, then queen can be represented by the equation Queen = King – Man + Woman (Stanford, n.d.). How do we get to that? Let's imagine words on a plot with dots representing each word as a datapoint. For the word "queen" we can imagine the datapoint would look like Figure 7.1.

Within each data point are the word's corresponding vectors, and the closer two words are together on the plot, the higher the probability that the words co-occur. Because "queen" would be closely related to the word "woman," the plot would look like Figure 7.2.

If you wanted to move from one data point to another, you could do so by subtracting or adding the distance between related words. This would mean that to find the location of "queen" on a graph, you would start at the location for "king," subtract the distance of "man," and add the distance of "woman" to land on "queen" (Figure 7.3).

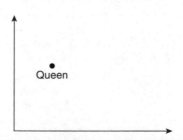

Figure 7.1   Plotting "queen."

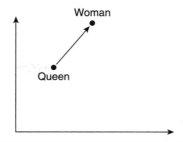

FIGURE 7.2   Plotting "queen" to "woman."

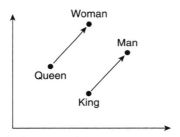

FIGURE 7.3   Plotting "queen" to "woman" and "king" to "man."

Within each datapoint are vectors that contain the probability of each word co-occurring with each other. When placed in a table, the resulting probabilities may look like Table 7.1.

As you can begin to see, these types of mathematical assumptions within the model, while very helpful in some contexts, can also cause some serious harm. Here are a few reasons:

1. While linearly mapping words can allow for easier mathematical representations, this is not quite the same as showing a model what a word actually means independent of its relationship with another word. Let's look at the equation, Queen = King – Man + Woman. This is more in line with a linear analogy than a definition of what the word "queen" actually means (Ethayarajh et al., 2018). On a graph, the analogy looks like Figure 7.4.

	k = Man	**k = Woman**
$P(k \mid \text{King})$	.875	.352
$P(k \mid \text{Queen})$	.303	.826

TABLE 7.1   Probabilities of Co-Occurrence King and Queen

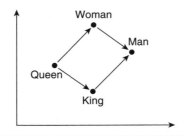

FIGURE 7.4 Plotting "queen," "woman," "king," and "man."

2. When words are turned into equations and scaled, words are associated in the model based on how often they co-occur with each other in the underlying dataset. This means that a biased dataset, like Common Crawl, carries within it a certain level of bias that can potentially be replicated if the resulting output from a GLoVe embedding model were to be scaled using a larger model—particularly one that uses weights.

CONSIDER THIS

Let's consider an example. If we multiply the probabilities of co-occurrences from a GLoVe embedding model with the weights within an attention layer like the ones used in models like BERT, we end up with a vector that contains the probability of "queen" and its probable co-occurrence with the word "king" as being less likely. In turn, words that are associated with the word "king," such as "man," are also multiplied by their probable co-occurrence with the word "queen" and its associates such as "woman."

Now consider what words may be further associated with "man" and "king" and what harms this can cause. If "king" is associated with "man" and "man" is associated with "strong," and "strong" is associated with "CEO" and "CEO" is associated with "president," what does it mean that "queen" is seen as "king's" opposite? It's clear that any model built on these assumptions will amplify bias. It's also evident that any model built using uncurated Internet data will multiply bias when this is the underlying source of its logic, regardless of what mathematical interventions you supply.

7.2.2 Section Summary

The underlying corpus of the GLoVe embedding model is the Internet, and LLMs that use self-attention layers such as BERT, RoBERTa, GPT-2, and GPT-3. These are popular language models that use data collected on the Internet with web crawlers (Common Crawl, n.d.b.). Researchers have shown that Internet data, particularly data taken from the United States and UK, have an overrepresentation of hegemonic views that encode biases potentially damaging to marginalized populations (Bender & Gebru, 2021).

Furthermore, researchers' interpretations of the capabilities of these types of models leave much to be desired. Unfortunately, there is a prevailing theory in NLP that the largeness of a dataset equates to the robustness of a model. This is not true. Common Crawl, the main public data source of GLoVE, RoBERTa, and GPT-3, among others, has petabytes of data from eight years of web crawling. However, the sourcing of this model can be particularly problematic (Common Crawl, n.d.b.). As Bender and Gebru (2021) point out, much of the Internet is skewed younger, and much of the Internet is skewed male, particularly on the main sites from which Common Crawl has sourced its data: Reddit and Wikipedia (Bender & Gebru, 2021). Additionally, websites like X (formerly Twitter), another GLoVE data source, have been proven to be quite biased by researchers such as Karen Spärck Jones (Bender & Gebru, 2021).

When we couple the lack of data curation with the faulty mathematical assumptions of these models, we find a recipe for harm that perpetuates damaging bias in LLMs.

PRACTICE

If you want to explore the power of word embeddings, head over to the open-source repository that serves as a companion to this chapter, https://github.com/jazmiahenry/fun-with-word-embs. There you will find a notebook called "word-embs-fun.ipynb" within the student_notebooks folder that you can use to get more familiar with GLoVe embeddings.

7.3 Natural Language Processing: Linear Decision Making for Nonlinear Language

Human behavior is inherently nonlinear. We may begin on one path making choices that may make it appear easy to predict our next move, but then, seemingly out of nowhere, we may pivot and make a choice that no one could have anticipated. While we are able to adapt to these changes in behavior as humans, computers are not as quick to change course. In popular NLP models, linearity is assumed, as numeric transformations of text are the source for LLM input. As discussed in the last section, we achieve this in NLP by using word embeddings created through vectorization, which linearly maps words to each other based on their Euclidean distance. By using this technique, we can mathematically quantify words in a way that machines can understand and build more robust language models without requiring extensive data labeling. While this approach offers many benefits, there are also potential downsides that must be considered.

Linear mapping can cause information loss in any nonlinear process, particularly when insufficient care is taken to minimize this loss. This issue is particularly relevant in the case of word embeddings because words do not always have perfect one-to-one mappings to other words. While words may share similarities, they can differ based on tone, context, and time.

For instance, consider the word "mad." In this model, it would typically co-occur or linearly map with the word "upset." However, in some contexts, "mad" and "upset" may not be similar at all. In the UK, "mad" can indicate someone experiencing a manic episode, while in the United States, some use "mad" to mean "really" or "very." In popular U.S. parlance, "mad" has also been used to signify "a large amount," as in the phrase "mad money." Thus, mapping the meaning of each vector and using matrix factorization to find its nearest neighbor will create a new representation of the word that cannot be perfectly mapped back to its original form.

When we increase the intensity of the strategy mentioned earlier through a noise robustness method called *stochasticity*, we observe an even greater amplification of bias in NLP. The stochastically robust process, which linearly maps words through self-attention layers in transformer models, aims to identify linguistic meanings that incorporate multiplied biases observed in simpler word embedding–based models. These models have been perceived to reduce the gap between human ability and machine comprehension, but they come with significant costs. These include increased replication of environmental harm, as models of this magnitude require additional computing resources, higher expenses due to increased computing, and more instances of gendered and racial bias (Bender & Gebru, 2021).

7.3.1   Attention Layer Mathematics

Like word embeddings, attention layers linearly map words and multiply them through a process of matrix multiplication that results in model output that helps advance a machine's ability to parrot language.

The mathematics of the self-attention layer are as follows:

$$\text{Attention}\,(Q, K, V) = \text{softmax}\left(\frac{QK^{T}}{\sqrt{d_k}}\right)V$$

where Q, K, and V stand for query, key, and value vectors, respectively.

Simply, the vanilla attention layer follows five steps (Figure 7.5):

1. A vanilla self-attention layer multiplies each input vector by weights that have been created during model training.

2. The resulting product is where you will extract the key, query, and value vectors discussed earlier.

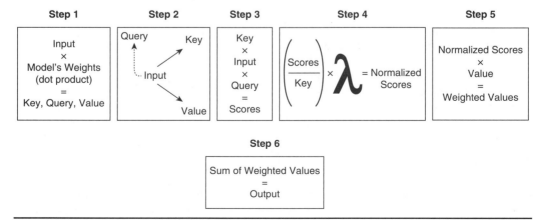

Figure 7.5   Attention layer mathematics illustration.

3. You then multiply your query vector by your current input and your other inputs against the key vector.

4. Next, you divide the resulting score by the dimensions of the key vector. You then regularize each score by applying the softmax function.

5. Lastly, you multiply the value vector by the result of the last step. Finally, you sum the weighted value of each vector, which will be the output for every word.

Think of the key, query, and value vectors as different representations of the same input. Say you want to find a list of comedy movies to watch online. The query vector is like inputting "best comedy movies" into your search bar, and the key vector would be a return of all of the keywords and tags of movies on the Internet that you are querying against. The value vector would be the return of information that matches the description given.

By multiplying the query vector with the key vector, you can find out which other inputs are most relevant to your query to find your value vector. The softmax function in the vanilla self-attention layer formula is simply a type of regularization function that converts your vectors into probabilities of K possible outcomes and normalized so that these scores add up to 1. This step ensures that you give more weight to the most relevant inputs while ignoring irrelevant ones.

Finally, you multiply the normalized scores with the value vectors to get the weighted values. These values represent how much information you should retrieve from each input to understand the current input better. By summing up these weighted values, you get the output for every word, which captures the most relevant information from all the inputs.

7.3.2   Section Summary: Outcome of the Attention Weights

These weights can be stochastic and add another layer of obscurity to what is already an opaque black box. There is nothing inherently problematic with this process—in fact, the pros of such a process are vast! One of the greatest advancements brought about from attention layers is the transformer. *Transformers* are deep learning models that use attention layers to quickly calculate the relationship between words in large datasets. The incorporation of the self-attention layer in models such as transformers have improved model performance natural language understanding, natural language generation, and generative AI tasks as well as work beyond traditional language modeling (Stanford, n.d.)! However, when the input to these models is already biased, multiplying by weights runs the risk of increasing the instances of biased language in the model's output. Not only that, but the linear mapping of the biased input bakes bias into the word's vectorized representations. This can mean that a word that may not on its own have a negative connotation, such as "Black" can have a higher probability of co-occurrence with negative words.

Therefore, it is important that you ensure that the mathematical assumptions you choose to support your model's creation do not multiply biases that harm marginalized folks. You can do this by:

1. Taking care to understand the assumptions your model is making and how it works in the real world

2. Being sure that the things you expect from the model (i.e., linguistic meaning) are in line with the actual abilities of the model (i.e., linguistic form)

3. Weigh the costs associated with opacity and make sure it is still worth pursuing this path

Now that we have worked through some examples of harms and ways they show up in our data and mathematical assumptions, let's get into ways to combat these harms.

PRACTICE

If you want to create word embeddings using transformers, head over to the open-source repository, https://github.com/jazmiahenry/fun-with-word-embs. There you will find a notebook called "tokenization-word-embs.ipynb" within the student_notebooks folder that you can use to get more familiar with using transformers to create word embeddings and visualize them.

7.4   Stage One: Data Collection

Most popular language models within the NLP field have used the open-source database called Common Crawl (n.d.a.). Common Crawl is an ultra-large database with petabytes of data that it claims represents over 40 languages and over 50 billion pages and trillions of links of data (Common Crawl, n.d.a.). Models such as BERT, RoBERTa, GPT-3, and GLoVe all use Common Crawl as its main data source.

As we discussed in a previous section, large databases run into the problem of properly documenting their data, leading to documentation debt (Bender & Gebru, 2021). In addition, when the databases get larger, it becomes harder and harder to begin documenting the data (Bender & Gebru, 2021). Thankfully, there are researchers who have been working on identifying the contents of Common Crawl and publicizing them. One site, called the Statistics of Common Monthly Archive, documents the representativeness of the data Common Crawl collects, the languages present in the data, and the size of each crawl (Common Crawl, n.d.b.). While the diversity of languages present in Common Crawl has improved since its inception, it is still heavily skewed toward English and more specifically toward US- and UK-based English. Forty-seven percent of Common Crawl as of this year is written in English. The second most popular language in the corpus is Russian at 5.5%, followed by German at 5.2% (Common Crawl, n.d.b.).

The quest for more representative data and combating documentation debt begin with more researchers being held accountable through data documentation. We do that by curating the data that we are training our models on so that we can identify where the gaps in the data are and work to fill them. We can do that through data transparency using tools that allow for data curation.

7.4.1   Dataset Nutrition Labels

In her book *Race After Technology,* Ruha Benjamin speaks about the importance of creating dataset nutrition labels that house not only the effectiveness of the curated data in our datasets but also the potential harms that might be present in them. This, much like the nutrition labels that we have on the back of our food, gives you a rundown of all your dataset's contents. There are many examples of data nutrition labels online that can be used as a tool for added data transparency. Authors at the Data Nutrition Project detail how to define data nutrition labels in their paper, "The Dataset Nutrition Label: A Framework to Drive Higher Data Quality Standards" (Holland et al., 2018).

7.4.2   Data Cards

The team at Google Research released a paper on *Data Cards*, a data transparency tool that documents information for your dataset such as how it was collected and when, the data annotation strategy used for the data, and how the data were used during training and model evaluation purposes (Pushkarna et al., 2022). These Data Cards can be kept alongside your model's code to allow researchers using your model to track information on your data's transformation and movement.

7.4.3   Data Documentation

It is important to save any changes you make to your model's features. Changes can include engineering new features from original features for the purpose of increasing model performance, using a dimensionality reduction technique that combines features of a large dataset into a smaller dataset with a reduction of features such as principal component analysis (PCA), transforming data points using imputation methods to remove nulls and outliers, and others. This documentation can be saved in a ReadMe file within a repository, by saving model artifacts in a model versioning or experimentation tool such as MLFlow, or with a data curation tool.

CONSIDER THIS

Regulatory reforms such as the General Data Protection Regulation (GDPR) requires increased data transparency of all personal data of EU citizens. This means collecting information on how data are stored and collected for auditing purposes.

While the United States does not have a national approach to data privacy protections, individual states do, and as of publishing, California, Colorado, Connecticut, Iowa, Utah, and Virginia all have data privacy protections that require transparency in how user data are collected.

It is important to learn how to properly document your data now to prepare for AI and data governance practices in the workplace that may require transparency depending on where the users you serve models to live.

7.5   Stage Two: Fight Bad Math with Better Math

Considering that the underlying problem of these models is not only rooted in their lack of data curation but also in the conflation between probability of occurrences with deriving meaning, it stands to reason that you can use probabilities to improve errancy in your model. This can be done in many ways.

7.5.1 Counterfactuals

Counterfactuals are a technique that comes out of the philosophy discipline that examines an alternative outcome for a model by questioning what would happen if a model's input were changed. In "Counterfactual Explanations Without Opening the Black Box," Wachter et al. describe counterfactual explanations as "a statement for how the world would have to be different for a desirable outcome to occur" (Wachter et al., 2016). Counterfactual explanations are useful for explaining why a certain decision was made by a model, and they can also be used to suggest changes to input features that would lead to a different decision. It can be helpful to use counterfactuals to add a layer of data transparency in your model when you want to understand the cause of resulting model predictions.

7.5.2 Parity

Parity metrics enable you to evaluate the instances of bias within a machine learning model, but in our case, can be used in NLP models as well (Barocas et al., 2023).

The first type of parity metric we are going to explore is *demographic parity* (Barocas et al., 2023). Demographic parity requires that the positive rates of the underrepresented subgroup be equal to the percentage of the positive rate of the overrepresented class. The probability equation of the outcome is defined as:

$$P(\hat{Y} \mid A = 0) = P(\hat{Y} \mid A = 1)$$

such that the positive rate of the overrepresented subgroup $a = 0$ is the same as the positive rate of the underrepresented subgroup $a = 1$. Using a confusion matrix, you can demonstrate the positive rate of each subgroup to visually prove demographic parity as well.

The second type of parity is called *equal opportunity parity*. This parity metric takes things a step further and says that, assuming all things are equal, the underrepresented subgroup should receive some prediction at the same rate as the overrepresented group. In this case, you would use the true positive rate within the confusion matrix for each subgroup to identify if the outcomes are equal. The formula of this probability is:

$$P(\hat{Y} = 1 \mid A = 0, Y = 1) = P(\hat{Y} = 1 \mid A = 1, Y = 1)$$

such that the true positive rate of the overrepresented subgroup $a = 0$ is the same as the positive rate of the underrepresented subgroup $a = 1$.

The last type of parity is the *equalized odds parity*. This metric is one that not only suggests that outcomes should be equal suggesting known factors remain constant but also that false positives across subgroups must be equal. To get the equalized odds parity rate, the machine learning practitioner would need to divide the true positive rate of each subgroup by the false positive rate for each subgroup. The goal is to make the percentage of the equalized odds parity metric as close to zero as possible. The probability formula is:

$$P(\hat{Y} = 1 \mid A = 0, Y = y) = P(\hat{Y} = 1 \mid A = 1, Y = y), \; y \in \{0, 1\}$$

such that the true positive rate of the overrepresented group $a = 0$ is equal to the true positive rate of the underrepresented subgroup $a = 1$ and the false positive rate of the overrepresented group $a = 0$ is equal to the false positive rate of the underrepresented subgroup $a = 1$.

7.5.3 Stratified Sampling

Stratified sampling is a technique to separate a population into groups by their significant similar characteristics. When separating groups by their demographics, you have the ability to apply stratified sampling to test potential dataset imbalance and model performance across groups.

When sampling across a population after using the stratified sampling technique, you can randomly choose a datapoint within a stratified sample to get a more true representative sample.

When testing model performance, stratified sampling allows you to run model scores across a demographic, run a correlation between the demography and outcome of a model, and even create categorical features that can serve as the basis of counterfactual models.

7.6 Stage Three: Model Constraints/Operations

Post model training and before model deployment, an important complementary script to your machine learning model includes a list of rules-based constraints that keep the model from diverging into the land of bias. These constraints or rules can be impeded in the main.py script, in the utils script, or in a script of its own. You want this script to bring with it a few main components:

1. You want the constraints to consider edge cases.

2. You want the constraints to be clearly written to reduce the likelihood of biased language reentering the model.

3. You want the constraints to be commented and documented for future users to understand the purpose of each constraint.

Constraints can be as restrictive or as loose as you decide, but it is better to have more simple constraints than difficult ones. It is best practice to follow the rule of clarity and abstraction when building a constraint and only add layers of complexity if necessary. So, for example, if you are building a constraint to, say, remove all instances of hateful language from a dataset, you would want to create straightforward rules that define to your model what hateful language is and document how you define that rule in your data's documentation for others to be aware of.

There are three flavors of constraints with different benefits for model incorporation that we will explore together.

7.6.1 Flagging

Flagging constraints are functions that return a flag every time a rule is breached. This can be as simple as returning "True" every time a word from a "banned list" is introduced into the model's lexicon or as complex as the incorporation of a couple for loops and while statements.

If creating a flagging mechanism in an object-oriented programing (OOP) language, your code would follow this logic: for every instance of a banned word, append this label to a list and return that list as a new feature that is the length of the dataset. This follows the same logic as creating a conditional and can be done using the "where," "select," or "if-else" statements in Python or the "if-else" statement in R. Flagging returns a boolean value as your output that you can incorporate into your dataset or designate specific values for added clarity. If you do not have structured data, you can flag words using annotation methods through data pipelines, JSON files, and YML files.

7.6.2   Pruning

Think of your language model as a rose and bias as its thorns. You can care for your rose through a process called pruning so that it can grow. You do that by removing the thorns. A pruning constraint is what happens when you remove biased instances completely from the model. This can be done either with or separately from any flagging action.

When coding this in an OOP language, your code will follow the logic: when a condition has been met, remove the word that fulfills the condition and remove it from the dataset. Flagging conditionals can be the foundation of the pruning method and serve as an intervention method for bias mitigation. This method can be paired with human-in-the-loop methods that allow for expert annotators to decide on a case-by-case method which words to remove. This process can be incorporated in the feature engineering process through data pipeline, JSON file, or human-in-the-loop system.

7.6.3   Nudging

The best way to discourage bad behavior of a model can be removing the bad from the model completely, calling attention to it through flagging, or ignoring it completely and encouraging "good" behavior in its place. Nudging is a way of calling attention to biased language in a dataset and nudging the model to make better choices. This can be done in a couple ways:

1. Replacing biased language with nonbiased language
2. After a model's output, downweighting negative associations so that better associations are ranked higher

In OOP, one can replace words with conditionals complete with "while" statements that replace harmful language, heuristics that gauge a model's weights and cut out negative word associations, and even prompt engineering to push a model to behave more appropriately. This method can be combined with flagging and pruning to improve model robustness and reduce instances of bias in the model. A popular method of nudging a model is using reinforcement learning from human feedback. This is a method where a researcher can rank a model's output by how complete it is to inspire the model to continue giving desirable output. This can be used to remove biased instances by ranking responses with less biased associations higher than models that have negative associations.

Now that we've gone through our three strategies to combat bias in NLP models, let's work together on a case study.

PRACTICE

Before moving on to the chapter's case study, head over to the open-source repository that serves as a companion to this chapter: https://github.com/jazmiahenry/fun-with-word-embs. There you will find a notebook called 'llm_training_tools.ipynb' within the student_notebooks folder where you can find helpful information on how to prepare training data for LLMs.

Case Study: The Limits of Better Training Data with No Constraint

In 2016, Microsoft released an online chatbot named Tay, a conversation AI bot that was deployed on Twitter with the purpose of increasing in intelligence and agility over time through conversations with people on the Internet (Vincent, 2016). It was trained by a publicly accessible dataset that used the Internet as its sole data source. Within 24 hours of his release, Tay went from being a jovial and happy bot to being an antisemitic, racist, and homophobic bot with an unsettling adoration of Hitler (Price, 2016) (Figure 7.6 and Figure 7.7).

While the model was not originally created to express harmful language, it was created with the intention of learning from the conversations with people on the Internet, and the "beauty, ugliness and cruelty [of the world]" began to shine through even as Microsoft wanted to focus only on its beauty (Benjamin, 2019).

1. How did this happen and how did researchers fall short of mitigating instances of bias with Tay?

2. What are some techniques that researchers could have used to document Tay's training data?

3. What are the mathematical assumptions of the model that may have led to some of the amplification of bias in Tay's output?

4. What are some constraint methods that the researchers can use to improve Tay's behavior in the future?

FIGURE 7.6 Four tweets of the robot Tay asking questions on X (formerly Twitter).

7.7 Chapter Summary

In this chapter, we discuss the issue of bias in LLMs that are trained using data pulled from the Internet, which often contain biases and perpetuate harmful stereotypes. This chapter begins by exploring how LLMs rely on linear mapping of word tokens into vectors, which become word embeddings that the models use to derive linguistic meaning. This approach can perpetuate biases that exist in the training data, as the models have an incentive to perpetuate these biases in their own sentence generation tasks.

To address this problem, this chapter suggests several solutions that have been divided into three stages: curating data sources through data labeling and data nutrition labels, weighing the effectiveness of data and ML models using parity metrics, and applying constraints and operations during the script-making process. This chapter describes how practitioners can make data-driven decisions at these three stages of model creation to improve model performance and reduce biased outcomes.

Overall, to overcome bias in LLMs, it is not enough to simply create supplementary corpuses and train them using pretrained models that have been built on harmful training data. Instead, practitioners must take active steps to curate their data sources and apply constraints and operations during the model-building process.

Chapter Glossary

Term	Definition
BERT	A pretrained model made by Google in 2018 that stands for "bidirectional encoder representations from transformers."
Common Crawl	An ultra-large database with petabytes of data that it claims to have representation of over 40 languages and over 50 billion of pages and trillions of links of data.

(Continued)

Term	Definition
Counterfactuals	A technique taken from philosophy, counterfactuals explore a model's "what ifs" by using alternative inputs to see the effect on a model's outcome.
Data card	A data transparency tool that documents information for your dataset such as how it was collected and when, the data annotation strategy used for the data, and how the data were used during training and model evaluation purposes.
Dataset nutrition label	Much like the nutrition labels that we have on the back of our food, this gives you a rundown of all your dataset's contents.
Demographic parity	Measures if a desirable outcome has an equal distribution across the majority group and all subgroups. So if the majority group has 45% acceptance to get a loan, the underrepresented subgroup should get 45% as well.
Equal opportunity parity	Measures if, all else being equal, the true positive rate of the majority group is the same as all subgroups. This means that if someone from the majority group A has a certain credit score and gets a loan, someone from underrepresented group B with the same credit score should also get the loan.
Equalized odds parity	Measures if the true positive rate of a majority group is equal to the underrepresented group with all known factors remaining constant and that false positives across subgroups must be equal to each other. The lower this metric, the more equal. In short, if someone from majority group A has a certain credit score and gets a loan and someone from underrepresented group B has the same credit score and gets a loan as well, but people with subpar credit scores in majority group A are more likely to be approved than underrepresented group B with subpar scores, then the parity metric will consider the model less equal.
GPT	A pretrained model made by OpenAI in 2018 that means generative pretrained transformer.
Linear analogy	Instead telling you the meaning of a word on its own, embeddings are maps showing how a word is related to another. In GLoVe embeddings can be defined as "a is to b as x is to y."
Log bilinear model	In language models, this is a function that computes a resulting context vector from previous vectors. It gives us the ability to predict a future word based on its relationship with a prior word, for example.
Matrix factorization	The process of multiplying matrices for a new resulting matrix.
RoBERTa	A pretrained model made by Facebook in 2019 that is an acronym for "a robustly optimized BERT pretraining approach."
Stochasticity	A process that refers to any output that has an element of randomness or probabilities or chance. The output of a stochastic process is hard to replicate even when you use the same input because the weights used to draw your output contain randomness.
Transformer	Deep learning model that uses attention layers to quickly calculate the relationship between words in large datasets.
Vectorization	A process of taking raw data such as the word "queen" and creating an array of vectors with real numbers showing its relationship with other words in the dataset.
Word embedding	Numeric representations of a word meaning stored in a vector.

References

Barocas, S., Hardt, M., & Narayanan, A. (2023). Fairness and Machine Learning: Limitations and Opportunities. MIT Press.

Bender, E., & Gebru, T. (2021, April). On the Dangers of Stochastic Parrots: Can Language Models Be Too Big? arXiv preprint arXiv:2104.04430. https://arxiv.org/abs/2104.04430

Benjamin, R. (2019). Race after Technology: Abolitionist Tools for the New Jim Code. Polity.

Common Crawl. (n.d.a.). Home. https://commoncrawl.org/

Common Crawl. (n.d.b.). Statistics of Common Crawl. https://commoncrawl.github.io/cc-crawl-statistics/

Ethayarajh, K., Duvenaud, D., & Hirst, G. (2018, October). Towards Understanding Linear Word Analogies. arXiv:1810.04882 [cs.CL]. http://arxiv.org/abs/1810.04882

Fraenkel, A. (2020). Fairness and Algorithmic Decision Making. https://afraenkel.github.io/fairness-book/intro.html

Holland, S., Hosny, A., Newman, S., Joseph, J., & Chmielinski, K. (2018). The Dataset Nutrition Label: A Framework to Drive Higher Data Quality Standards. CoRR, abs/1805.03677. http://arxiv.org/abs/1805.03677

Noble, S. (2018). Algorithms of Oppression: How Search Engines Reinforce Racism. New York University Press.

Price, R. (2016, March 24). Microsoft Deletes Racist Genocidal Tweets from AI Chatbot Tay. Business Insider. https://www.businessinsider.com/microsoft-deletes-racist-genocidal-tweets-from-ai-chatbot-tay-2016-3?r=UK&IR=T

Pushkarna, M., Zaldivar, A., & Kjartansson, O. (2022, April). Data Cards: Purposeful and Transparent Dataset Documentation for Responsible AI. arXiv preprint arXiv:2204.01075.

Stanford. (n.d.). GLoVe: Global Vectors for Word Representation. https://nlp.stanford.edu/projects/glove/

Vincent, J. (2016, March 24). Microsoft Chatbot Racist. The Verge. https://www.theverge.com/2016/3/24/11297050/tay-microsoft-chatbot-racist

Wachter, S. (2017, November 1). Counterfactual Explanations without Opening the Black Box: Automated Decisions and the GDPR. arXiv.org. https://arxiv.org/abs/1711.00399

End of Chapter Questions

1. What are some ways that bias can show up in a large language model?

2. Name three ways to reduce bias in a large language model.

3. What are the mathematical assumptions of word embedding models such as GLoVe?

4. How do the assumptions of word embeddings affect model reasoning and large language model output?

5. Where is much of the data being used to train large language models sourced?

6. How can data documentation be used to improve a large language model?

7. What are transformers and how have they been used to advance the technology of large language models?

8. What are three types of model parity and how are they used to measure bias in a large language model?

9. Describe the steps of attention layers. How are they used in large language models?

10. Now that you have information on how to process and visualize word embeddings as well as how to process your data to train large language models responsibly, how would you apply this to the real world?

CHAPTER 8

Recognizing Bias in Medical Machine Learning and AI Models

Isaac K. Gang

George Mason University

Question: How can you mitigate bias in medical machine learning and AI systems?

Learning Objectives

Upon completion of this chapter, the student should be able to

- Explain the history of machine learning
- Define machine learning
- Build a simple machine learning model
- Discover health care bias and inequities in machine learning

Chapter Overview

In this chapter, we discuss basic machine learning concepts and bias in medical machine learning and artificial intelligence (AI) systems.

8.1   Introduction

Ready, set, machine learning! When Arthur Samuel of IBM coined the term "machine learning" in 1959, the computing community expected nothing more than another addition of a fancy research term to the plethora of terms that were already in existence. In fact, not many—unless you were into computer board games—were aware of his work. Samuel's research, entitled "Some Studies in Machine Learning Using the Game of Checkers," was profound because, for the first time, it allowed us to discover that a machine can defeat the person who programmed it (Samuel 1959, 535). Considered the father of machine learning, perhaps arguably, Samuel successfully opened the door to this exciting, yet sometimes controversial, topic that will continue to shape the future of humans for years to come. In fact, the steady popularity of machine learning as a sub-field of computer science, with noticeable explosion in 2004, was observed in many scholarly publications and presentations (Theobald, 2017, 8). Figure 8.1 shows the historical mentions of "machine learning" in published books.

Furthermore, thanks to the power of machine learning, the way we solve problems has been forever changed because we can now do more with less. Specifically, data and programming have been intertwined, and this allows people, even those with little to no programming experience, to join in the fun and solve problems conveniently, cheaply, and efficiently. We formally define machine learning in the next section.

8.2   Defining Machine Learning

Machine learning is straightforward because the name sort of gives it away. In short, machine learning is an automated process that extracts patterns from data (D'Arcy et al., 2015, 3). In other words, we are able to train the machine, and it is able to learn from data.

Though machine learning is a great discovery, a timely advance in the fields of AI and computer science, it is not, however, without its issues, chief among which is bias. The secret weapon to machine learning is data, and because data are growing rapidly, machine learning continuously evolves with it. Two of the main categories are supervised and unsupervised learning. To build a machine learning model, it must be identified as either supervised or unsupervised, with supervised being the most common. With supervised, data scientists or programmers program expertise to learn from data (Deitel & Deitel 2019, 596). The two subcategories of machine learning are defined next.

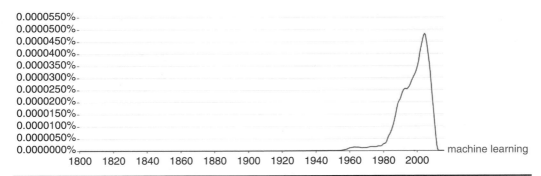

FIGURE 8.1   Historical mentions of "machine learning" in published books (Google Ngram Viewer, 2022).

8.2.1   Supervised Machine Learning

Supervised machine learning implies the presence of a supervisor that supervises the machine. In other words, the machine is trained with labeled data.

The presence of data (or a dataset) is key, or the machine cannot learn. In this context, a data row is often referred to as a sample and a data column as a feature of that sample. Each sample has a label or target, which is the value we want to predict. Supervised machine learning can be further refined into **classification** and **regression**. In classification, the model attempts to predict discrete classes or categories to which samples belong. In the case of binary (two classes) classification, a sample is either part of a given class or it is not. In an email classification, for example, a sample (an email in this case) is either a "spam" or "not spam" (Deitel & Deitel 2019, 597). In regression, the model attempts to predict a continuous output to which samples belong. If we are interested in predicting the weather or stock exchange, for example, we would use regression.

8.2.2   Unsupervised Machine Learning

Unsupervised machine learning implies the absence of a supervisor that supervises the machine. In other words, the machine is trained with unlabeled data. So, without supervision or guidance, the machine is able to group uncategorized information according to similarities, patterns, and differences without any prior training of data. It therefore uses clustering algorithms like k-means, among others, to make the prediction.

8.3   Building a Simple Machine Learning Model: Use Case 1

In order to get a feel for the terse machine learning coverage discussed earlier, we will build a simple machine learning model as our first use case. The second and last use case will investigate bias and inequity in health care when machine learning is applied by looking at a health care dataset. Both of these use cases are meant to get your hands a little dirty, which is the best way to understand the fundamentals behind machine learning as a topic and dataset manipulation as a data analytics process.

8.3.1   Preparing Our Development Environment

As an artist needs a canvas to draw, a data scientist needs an integrated development environment (IDE) to write code and build models. In this section, we will go through the steps to download Anaconda, one of the most popular data science canvases on the market. Following these steps patiently and carefully will allow you to build a model in no time, even with minimum or no programming experience. Similar or related steps or processes can be found in (Deitel & Deitel, 2019), (Theobald, 2018a; Theobald, 2018b), and (D'Arcy et al., 2015, 3). In order for you to run the following commands successfully, you need to be able to access your terminal. While accessing the terminal varies by computer, the following steps allow you to open your terminal on macOS and Windows platforms.

8.3.2   Downloading Anaconda by Going to www.anaconda.com/download

Then access the terminal by using the following steps:

- Applications → Utilities (macOS)
- Start Menu → Anaconda prompt (Windows)

These steps and commands assume that you are running your system as the administrator and you are working in the current folder.

8.3.3   Updating Anaconda Packages from Terminals

- Conda update conda
- Conda update –all

8.3.4   Installing the Prospector Static Code Analysis Tool

- pip install prospector

8.3.5   Installing jupyter-matplotlib Visualization Library: ipympl

- conda install -c conda-forge ipympl
- conda install nodejs
- jupyter labextension install @jupyter-widgets/jupyterlab-manager
- jupyter labextension install jupyter-matplotlib

Once these steps and commands are completed successfully, you should be able to work in either ipython (interactive mode) or Jupyter Notebook. Note: Python programming has two modes from which you can work—these are interactive mode, where you enter commands and see results instantly, and script mode, where you write a program and run it once done. The following commands launch ipython and Jupyter Notebook, respectively:

- With your terminal opened, type "ipython" (without the quotes) to enter the interactive mode.
- With your terminal opened, type "jupyter lab" (without the quotes) to enter Jupyter Notebook's interactive and script modes.

Examples in this chapter were completed in the ipython interactive mode.

8.3.6   Standard Steps to Build a Machine Learning Model

The following are standard steps to follow when building a machine learning model (Deitel & Deitel, 2019, 599):

1. Load the dataset
2. Explore the dataset
3. Transform your data
4. Split the dataset for training and testing
5. Implement the model
6. Train and test the model
7. Run and evaluate the accuracy of the model
8. Make a prediction

8.3.7   Classifying Digits with the k-Nearest Neighbor Algorithm

Scikit-Learn is one of the most popular machine learning libraries available, and it comes with built-in datasets. In this example, we will work with the digits datasets similar to what is done by Deitel and Deitel (2019, 599). Code demo may be found in textbook supplemental material.

8.4   Health Care Bias and Inequities: Use Case 2

In your study of this book, you will come to learn two important terms that are crucial in machine learning and AI. These are **overfitting** and **underfitting**. In an effort to create more balanced and generic machine learning models, data scientists and analysts have to make sure that their model is generalized. That is to say they work the same under varying conditions. A model that doesn't generalize is considered overfitted, while those that do a little too well are underfitted. Too much of either is undesirable (The Health Inequality Project, n.d.).

8.4.1   Background

The saying "like a black swan" is an old expression that originated from Europe in the 17th century in reference to something that is impossible. The black swan theory, as we know now, is among the first known cases of bias in humans. The Europeans at the time believed that all swans were white. It wasn't until a Dutch explorer by the name of Willem de Vlamingh discovered black swans in Australia that this belief came to an abrupt end. While one might be tempted to question the open-mindedness of those who held this belief, it is perfectly normal for any of us humans to hold such beliefs given the circumstances. The only swans that these Europeans encountered and observed at the time were white. So, their mental model—that is, their mindset shaped by their environment—was molded by this reality, which is why they didn't think there could possibly be a swan of another color anywhere. This is implicit bias. If we were to drop machine learning and AI on their laps and ask them to build a swan model, it would have been biased toward a white swan—that is, the model would only recognize a white swan. If you feed it a black swan, it will not recognize it as a swan. This is a perfect example of an overfitted model. While this is not a desired outcome for a machine learning model, you also do not want a model that classified everything it sees as a swan regardless of what it is. As you might have guessed, that would be an example of an underfitted model. Our friends from that period have definitely failed the rule of verifiability discussed by Hakan (2021, 1655).

8.4.2   Are We There Yet?

The obvious question would be how much progress have we made since? Put another way, are we there yet? The short answer is no. What follows is the long answer. Machine learning has the potential to transform our lives in positive ways if we understand the implication and effect of bias in data, algorithms, and resulting systems as described earlier. A lot of times, when the word bias is mentioned, the assumption is made that we are talking about explicit bias— where someone is intentional about his or her actions. And because the person is often very clear about his or her feelings and attitudes, the related behaviors are conducted with intent. Explicit bias is processed neurologically at a conscious level as declarative, semantic memory and in words. It then manifests itself via negative behavior generally expressed through discrimination, physical and verbal harassment, or exclusion. These characteristics can, in and of themselves, serve as a deterrent because they are easily recognizable. That is why assumption of explicit bias is incorrect because such bias is clearly an exception, not the rule. The most dangerous bias—the kind we really need to watch out for—is implicit bias. It is dangerous because we don't do it intentionally because our environment (refer to the black swan theory) and upbringing shape it and, thus, bias operates outside of the person's awareness. As such, it is generally in direct contradiction to an individual's core beliefs and values. What makes this type of bias dangerous is that it automatically creeps into a person's thoughts and actions without their knowledge. It therefore affects important decisions. This is the kind of bias we struggle with in academia. As such, we need to intentionally combat this kind of bias in all we do if we are to create an environment that is truly diverse, equitable, and inclusive.

8.4.3   The Use Case

This use case investigates cases of biased systems and their impact on society and implements a machine learning model that showcases inherent bias as the result of including or excluding specific features and variables. It will further attempt to make the connection between these biased scenarios, on one hand, and inequity (disparity in rates due to differences in social, economic, environmental, or health care resources), inequality (varying rates with the amount of the resource and uneven distribution of resources among resource groups), disparity (difference in health status rates between population groups), and burden (how people are affected in specific groups and in the total population), as defined by Klein and Huang (n.d.), on the other. The area of health care continues to lag behind and suffer from bias, both explicit and implicit, in machine learning and AI scholarship, models, and systems. Health care bias is long-standing problem that has since been exacerbated by the increased use of machine learning and AI in treatment and care decisions and referrals.

8.4.3.1   Comorbidity

This model (with description and underlying implementation due to space constraint) is based on commodity—a condition describing the presence of more than one disorder in a person at a given time. In this use case, any combination of diseases could have been a candidate, but the focus is placed on cancer and mental illness, conditions that are common within the underrepresented minorities (URM) population. And since there are various types of cancer, this use case will specifically focus on lung cancer, a variance of which has been investigated by data analytics students at George Mason University in collaboration with the Allwyn Corporation. Also, several studies, including Klein and Huang (n.d.), National Institute of Child Health and Human Development (n.d.), Department of Health (n.d.), and The Health Inequality Project (n.d.), have shown that those with severe mental illness (SMI) frequently have worse outcomes than those without. But these studies do not examine such variables as race, ZIP code, age, and economic status—these are critical factors within the URM populations, and this is the reason this model specifically prioritizes and investigates these variables. Building on these findings and using the SMI diagnosis type combined with patient and hospital data as factors for predicting outcomes, it is possible to find areas where treatments and referrals can be improved for those in the target group (URM).

In this model, supervised machine learning techniques are used throughout the process, with random forest, XGboost, and logistic regression used predominantly on the Healthcare Cost & Utilization Project (H.CUP) (Department of Health, n.d.) datasets.

8.4.4   Implementation Results and Brief Explanation

The results in Figures 8.2 to 8.7 show clear bias in most of the categories examined. Each figure is accompanied by a brief description of what is happening with the related machine learning implementation behind the scene.

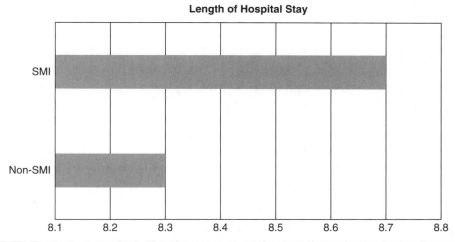

FIGURE 8.2   The mean length of stay in days for patients with severe mental illness (SMI) compared to patients without.

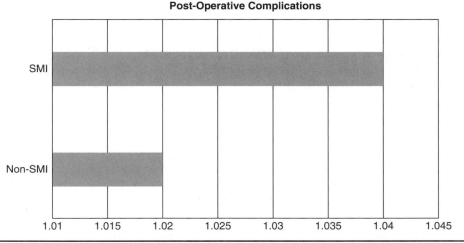

FIGURE 8.3   The post-operation complications data for patients with SMI versus without SMI.

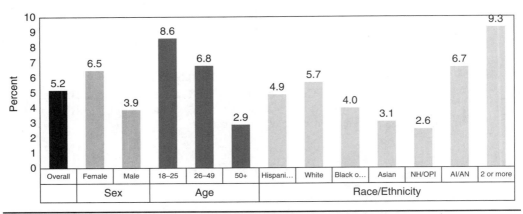

FIGURE 8.4 The prevalence of serious mental illness among U.S. adults by sex, age, and race.

Mental Illness Groupings	Groupings Description	% Deceased	Avg LOS	Avg Cost of Stay
F01-F09	Mental disorders due to known physiological conditions	4.04%	12.44	$184,709.99
F10-F19	Mental and behavioral disorders due to psychoactive substance use	1.58%	7.63	$119,313.94
F20-F29	Schizophrenia, schzotypal, delusional, and other non-mood psychotic disorders	3.30%	9.42	$162,030.66
F30-F39	Mood [affective] disorders	1.34%	7.56	$117,407.65
F40-F49	Anxiety, dissociative, stress-related, somatoform and other nonpsychotic mental disorders	1.38%	7.41	$117,194.48
F50-F59	Behavioral syndromes associated with physiological disturbances and physical factors	0.00%	10.81	$156,455.62
F60-F69	Disorders of adult personality and behavior	0.00%	7.43	$120,852.29
F70-F79	Intellectual disabilites	22.22%	13.44	$153,039.89
F80-F89	Pervasive and specific development disorders	0.00%	11.60	$128,718.40
F90-F98	Behavioral and emotional disorders with onset usually occuring in childhood and adolesence	0.00%	8.17	$125,469.43
F99	Unspecified mental disorders.	0.00%	2.50	$153,035.50

FIGURE 8.5 Patient outcomes by SMI grouping category.

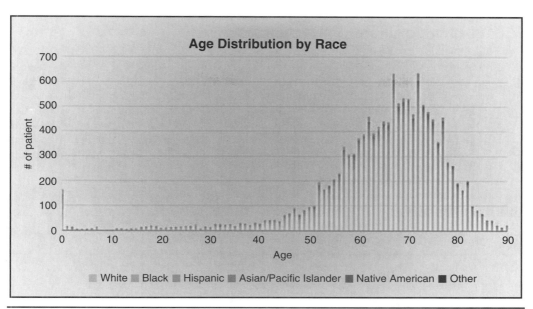

FIGURE 8.6   A histogram color-coded by race indicating frequency of patient ages for all lung cancer patients.

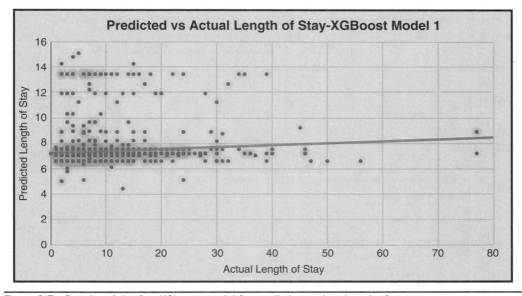

FIGURE 8.7   Results of the first XGboost model for predicting patient length of stay.

8.5   Chapter Summary

Bias has been established to be a serious problem in computing systems. In particular, it has been shown that the outcomes of many machine learning and AI-based systems can be subject to systematic errors in their ability to classify subgroups of patients, estimate risk levels, or make predictions. Because these errors can be introduced across the various stages of development, decisions with respect to algorithms and accompanying datasets should be carefully reviewed and examined before use. For example, this and other works demonstrated that the application of a commercial prediction algorithm can result in significant racial bias in predicting outcomes of comorbidity and other medical conditions.

Chapter Glossary

Term	Definition
Classification	The model attempts to predict discrete classes or categories to which samples belong.
Commodity	A condition describing the presence of more than one disorder in a person at a given time.
Machine learning	An automated process that extracts patterns from data.
Overfitting	A model that is not generalizable.
Regression	The model attempts to predict a continuous output to which samples belong.
Supervised machine learning	Has the presence of a supervisor that supervises the machine. In other words, the machine is trained with labeled data.
Underfitting	A model that is too generalizable.
Unsupervised machine learning	Implies the absence of a supervisor that supervises the machine. In other words, the machine is trained with unlabeled data.

References

D'Arcy, A., Mac Namee, B., & Kelleher, J. D. (2015). Fundamentals of Machine Learning for Predictive Data Analytics: Algorithms, Worked Examples, and Case Studies. MIT Press.

Deitel, P., & Deitel, H. (2019). Intro to Python for Computer Science and Data Science: Learning to Program with AI, Big Data and The Cloud. Pearson Education.

Department of Health. (n.d.). Comorbidities A Framework of Principles for System-Wide Action. https://assets.publishing.service.gov.uk/government/uploads/system/uploads/attachment_data/file/307143/Comorbidities_framework.pdf

Google Ngram Viewer. (n.d.). In Google Books Ngram Viewer. Retrieved May 10, 2024, from https://books.google.com/ngrams/ for historical mentions of machine learning in published books.

Hakan, T. (2021). Philosophy of science and black swan. Child's Nervous System, 38,1655–1657. https://doi.org/10.1007/s00381-020-05009-3

Klein, R., & Huang, D. (n.d.). Defining and Measuring Disparities, Inequities, and Inequalities in the Healthy People Initiative. https://www.cdc.gov/nchs/ppt/nchs2010/41_klein.pdf

National Institute of Child Health and Human Development. (n.d.). Health Disparities: Bridging the Gap. https://www.nichd.nih.gov/sites/default/files/publications/pubs/documents/health_disparities.pdf

Samuel, A. L. (1959). Some studies in machine learning using the game of checkers. IBM Journal, 3(3), 535–554.

The Health Inequality Project. (n.d.). How Can We Reduce Disparities in Health? https://healthinequality.org/

Theobald, O. (2017). Machine Learning for Absolute Beginners: A Plain English Introduction. Scatterplot Press.

Theobald, O. (2018a). Machine Learning: Make Your Own Recommender System. Amazon Digital Services.

Theobald, O. (2018b). Statistics for Absolute Beginners: A Plain English Introduction. Scatterplot Press.

Theobald, O. (2019). Machine Learning with Python: A Practical Beginners' Guide. Amazon Digital Services.

End of Chapter Questions

1. What is comorbidity? Give two examples.

2. Define machine learning and provide four real-life applications that used commercially today.

3. Recidivism in criminal justice is a real problem. What does it mean and how can you minimize, if not mitigate, it?

4. There are two types of bias. What are they?

5. Describe one example of bias that continues to persist.

6. Perform a Google search to find out the height of a car crash dummy. What is it and what are the bias implications?

CHAPTER 9

Toward Rectification of Machine Learning Bias in Health Care Diagnostics: A Case Study of Detecting Skin Cancer Across Diverse Ethnic Groups

Jennafer Shae Roberts

Researcher, Accel AI Institute

Laura Naomi Montoya

Executive Director, Accel AI Institute

Question: How can racial bias be mitigated in skin cancer detection algorithms?

Learning Objectives

Upon completion of this chapter, the student should be able to

- Gain a comprehensive understanding of the risks associated with biases in technological health care applications

- Understand the significance of mitigating harms in machine learning (ML) applications used at scale in health care, where they have a direct impact on people's lives and well-being
- Explain how biases can impact decision making in health care technology and the potential consequences
- Identify and describe the following biases and how they apply to ML applications for health care: cognitive, evaluation, sampling, underestimation, and statistical biases
- Understand where to implement bias mitigation strategies in the ML life cycle including in preprocessing, in-processing and postprocessing phases

Chapter Overview

The first part of this chapter covers a case study of skin color bias in image recognition for melanoma, the deadliest form of skin cancer (American Academy of Dermatology, 2016). The datasets used for training had previously only included light-skinned images, excluding populations with dark skin and creating a bias in the algorithm (Daneshjou et al., 2021). Because ML can now be used to detect melanoma with similar accuracy to that of board-certified dermatologists, it is important to learn how biases such as this occur and how they can be mitigated. ML has the potential to exacerbate health care disparities if it is not built with inclusivity in mind (Adamson & Smith, 2018).

9.1   Introduction

Say you discover a suspicious mole or lesion on your skin, and instead of getting an expensive and invasive biopsy done, you could take a picture with your smartphone, and a diagnostic app that uses ML algorithms would tell you if it was cancerous or not. But there's a catch: it only works if you are white.

Diagnostic ML tools such as those for detecting melanoma skin cancer have great potential. But shouldn't these technologies be available to everyone, not just to those with light-colored skin? It is important to recognize and mitigate bias in ML algorithms to address how historical bias is reproduced by algorithms and how to prevent bias at different stages.

9.1.1   How Does ML Bias Occur, and How Do We Mitigate It?

Bias in ML has many real-world consequences and has the potential to cause substantial harm (Adamson & Smith, 2018). How algorithms make decisions differs from humans in two key ways:

1. Algorithms are trained or developed to accomplish very narrow tasks.
2. Algorithms can be black boxes that lack interpretability and transparency.

Due to this, a model should not be evaluated on the same dataset it was trained on, but it is usually evaluated on a similar dataset representing the same type of population, as depicted in Figure 9.1. Data are generated from a subset of the relevant population, which then undergoes training and testing for the ML model. The predictions the model makes then trigger diagnosing decisions and actions, which are applied to the larger

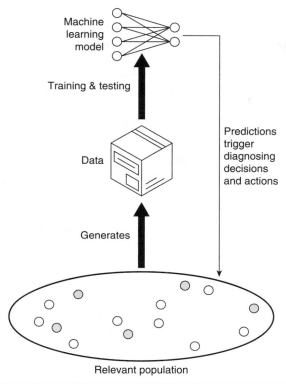

Machine
learning
model

Training & testing

Data

Generates

Predictions
trigger
diagnosing
decisions
and actions

Relevant population

FIGURE 9.1   Population data feedback loop.

(Inspired by van Giffen et al., 2022.)

population and may not be relevant to everyone who was outside of the training pool. As an effect of this system, the predictors may be inaccurate for those individuals who are outside of the training set.

If the benchmark, which is what model outcomes are compared against, is not representative in itself, preference would be given to models that perform well on a subset of the relevant population (Suresh et al., 2018). There is a danger of overlooking potential biases if the wrong benchmark is chosen. Mitigation strategies include the following:

- Monitoring for bias throughout the life cycle of the ML application
- Making ML models as transparent and explainable as possible (Van Giffen et al., 2022)

9.2   Case Study: Mitigating Bias in ML for Melanoma

When addressing bias in ML, it is imperative to address ML use cases in health care, which can influence life and death decisions. Bias in health care, as well as in datasets, can cause disparities in who benefits from the technology and who experiences marginalization. Within medical research, the term **bias** refers to "a feature of the design of a study, or the execution of a study, or the analysis of the data from a study, that makes evidence misleading" (Pot et al., 2021; Stegenga, 2018).

Rumor debunked: Is melanoma a white-skinned disease? Although light-skinned populations are at higher risk for melanoma, dark-skinned communities are at higher risk of death from this disease (American Academy of Dermatology, 2016). Early detection is key, as it greatly increases survival rates. ML melanoma detection technology must be bias free and accurate for people across all skin tones. Can AI help? Figure 9.2 shows AI-generated images of skin cancer on dark skin tones, which could possibly help to train the algorithm without using invasive and privacy-breaching images from people of color with melanoma.

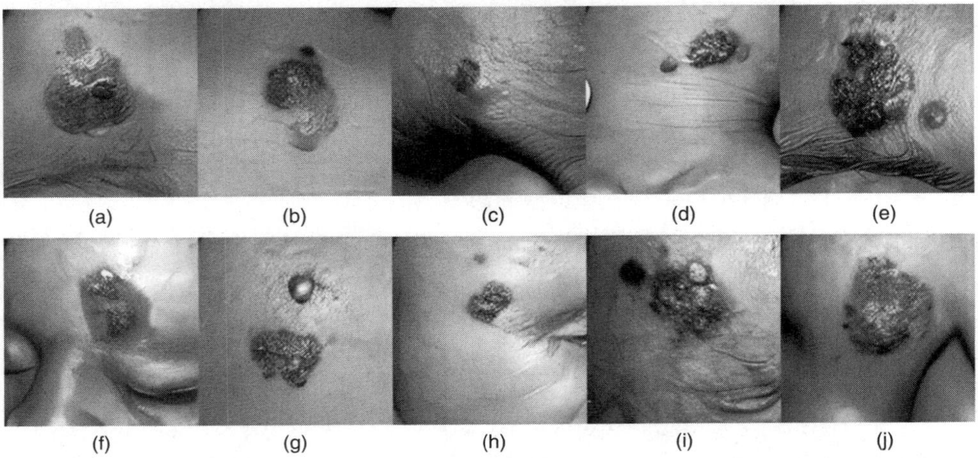

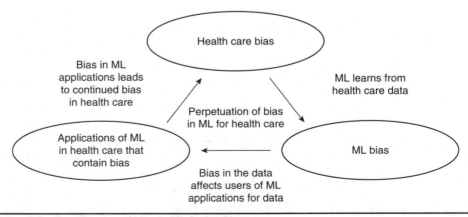

(a)   (b)   (c)   (d)   (e)

(f)   (g)   (h)   (i)   (j)

Figure 9.2   AI-generated images of melanoma on skin of color.

(Kvak et al., 2023)

Figure 9.3 demonstrates the perpetual nature of bias in health care, such as the racial bias in melanoma detection algorithms. Because ML learns from health care data that have existing bias, ML adopts the bias. Bias gets built into applications of ML that are used for health care, and the bias continues. Thus, we see the perpetual nature of bias and how it can be reproduced if nothing is done to address it.

Health care bias

Bias in ML applications leads to continued bias in health care

ML learns from health care data

Perpetuation of bias in ML for health care

Applications of ML in health care that contain bias

ML bias

Bias in the data affects users of ML applications for data

Figure 9.3   Perpetuation of bias in ML for health care.

Melanoma is traditionally diagnosed by a dermatologist using the ABCDE procedure characteristics, distinguishing malignant melanomas from benign skin lesions (Figure 9.2):

A. Asymmetry

B. Border irregularity

C. Color variability

D. Diameter greater than 6 mm

E. Evolution or any kind of change over time (Daghrir et al., 2020)

In order to diagnose skin lesions at the earliest possible stage, computer vision algorithms are being developed (Thomsen et al., 2020). Methods of classification are varied and include decision trees (DTs), support vector machines (SVMs), and artificial neural networks (ANNs; Dhivyaa et al., 2020; Hekler et al., 2019; Kaur et al., 2022; Murugan et al., 2019).

Dermatologists and patients can benefit from digital image processing, which helps in diagnosing skin lesions without making any physical contact with the skin (Daghrir et al., 2020). Processes of ML melanoma detection include (Figure 9.5):

1. Image acquisition and preprocessing

2. Lesion segmentation

3. Lesion characterization

4. Lesion classification (Daghrir et al., 2020)

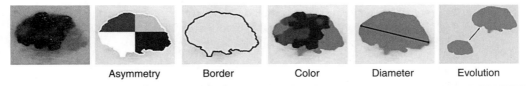

| Asymmetry | Border | Color | Diameter | Evolution |

FIGURE 9.4   ABCDE for melanoma diagnosis.

(Kvak et al., 2023)

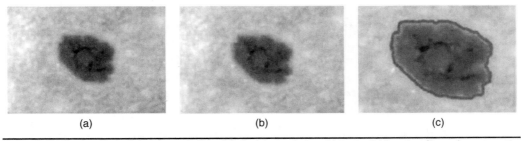

(a)   (b)   (c)

FIGURE 9.5   Computer lesion segmentation: (a) dermoscopic input image, (b) image after enhancement, and (c) segmented lesion.

(Jaworek-Korjakowska, 2016)

A critical step in melanoma detection after the image is taken is lesion segmentation, which isolates the pathological skin lesions from the surrounding healthy skin. ML techniques filter the image with the gradient magnitude in order to separate the lesion in the foreground from the background (Daghrir et al., 2020). Color labels are utilized to detect the presence of black and blue colors in a lesion, based on the color name (CN) features, which include 11 different linguistic color labels. Physicians recognize melanoma by detecting blue-black colors, where if at least 10 percent of the surface of a lesion is blue or black pigmented, it indicates melanoma (Daghrir et al., 2020; Kato et al., 2019). However, for an algorithm, the error rate will increase in the detection of lesions on dark-skinned patients if the training data only included images of lesions on light skin, thus leading to sampling bias. The ABCDE and the blue-black rule are the two systems that are relied on by most health care workers for melanoma detection, with obvious limitations often skewed by cognitive bias and which can lead to sampling and evaluation biases, all described in Section 3 (Daghrir et al., 2020).

9.2.1 Retraining Melanoma Detection Algorithms for Diverse Skin Tones

If nothing is done to correct bias, it will continue to be marginalizing. Figure 9.6 demonstrates how the case study of melanoma detection algorithms exemplifies the risks of continued bias with inaction. Similar to Figure 9.3, we see a perpetual cycle of bias, where racial bias in health care influences medical studies and datasets which feature only white-skinned people and cause a lack of diverse skin tones in dermatology datasets. This lowers the accuracy for anyone who doesn't have light skin, which creates an algorithm that only detects skin cancer on light skin. This leaves dark-skinned individuals marginalized and perpetuates a cycle of racial bias in health care, requiring bias mitigation.

Regardless of race or skin tone, anyone can develop melanoma, and early detection can be lifesaving. Hence, a technology that could aid in early detection would be beneficial to everyone. The general stages of melanoma skin cancer are described in Table 9.1.

Rates of the severity of illness and risk of death are increased in people with darker skin. Although overall rates of diagnosis are lower, the disease is more severe. Skin types are described by the Fitzpatrick skin type scale (Fitzpatrick, 1975), shown in Figure 9.7, which also illustrates how although the risk of melanoma is less for dark skin, the risk of mortality from melanoma is higher.

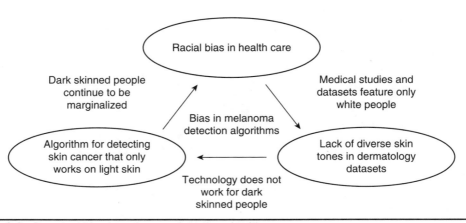

FIGURE 9.6 Perpetual bias in melanoma detection algorithms.

Stage of Melanoma	Survival Rate
Stage 1: in situ	99%
Stage 2: high-risk level	45%–79%
Stage 3: regional metastasis	24%–30%
Stage 4: distant metastasis	7%–19%

TABLE 9.1 General Stages of Melanoma and Survival Rates (World Cancer Research Fund, 2022)

Skin type	Skin reaction to sun exposure	Melanoma risk	Lethality of melanoma
I	Always burn never tan	High	Low
II	Burns easily tans minimally		
III	Burns moderatelly tans gradually to light brown		
IV	Burns minimally tans to moderatelly brown		
V	Rarely burns tans easily		
VI	Never burns dark pigmentation	Low	High

FIGURE 9.7 The Fitzpatrick skin typing test.
(inspired by Micu, 2014 and Hu et al., 2006)

The Fitzpatrick skin typing test was developed for dermatology to study human skin pigmentation in response to ultraviolet (UV) light. This scale includes six categories ranging from the palest to the darkest brown skin tones.

9.2.2 The Diverse Dermatology Images Dataset

The images in the **Diverse Dermatology Images (DDI) dataset** include a sampling across all Fitzpatrick I to VI skin tones for direct comparison of testing on images labeled with the lightest I/II (control group) and darkest V/VI skin tone categories. The DDI dataset was curated in 2022 as the first of its kind which represents diverse skin tones in a publicly available and pathologically confirmed dataset. This is a step toward allowing algorithms to aid in triaging skin diseases effectively and inclusively. There were no public artificial intelligence (AI) benchmarks that contained images of biopsy-proven malignancy on dark skin before the DDI dataset (Daneshjou et al., 2021). Frequently used datasets for melanoma detection, including ISIC (Rotemberg et al., 2021), PH2 (Mendonca et al., 2013), and HAM10000 (Tschandl, 2018), do not contain metadata for ethnicity or skin tone descriptors. In 2021, a review was done of 70 AI studies for dermatology titled "Lack of Transparency and Potential Bias in Artificial Intelligence Data Sets and Algorithms." (Daneshjou et al., 2021). Most publications were missing information on patient skin tone, race, or ethnicity, and the dataset and models had not been made public (Daneshjou et al., 2021).

The DDI dataset includes a retrospective of 656 images representing 570 unique patients with lesions of interest gathered from Stanford Clinics pathology reports over a 10-year period (Daneshjou et al., 2021). The lesions were classified as benign or malignant and include relevant labels for age, gender, and Fitzpatrick skin type of the patient. The labeling for lesion diagnosis as well as the skin tone classifications was cross-verified by patient follow-up records and image reviews by board-certified dermatologists.

The DDI dataset was tested against three separate state-of-the-art AI algorithms trained to classify lesions in images as benign or malignant (Esteva et al., 2017; Han et al., 2020; Tschandl, 2018). The methodology for training and testing by the Stanford researchers is described in Daneshjou et al. (2021). Their code was made publicly available for further research and testing purposes (DDI Alliance, 2022).

In the study, the three algorithms (ModelDerm, DeepDerm, and HAM 10000) were tested on the DDI dataset (Daneshjou et al., 2021). When comparing the subsets of Fitzpatrick I to II and V to VI images, all three algorithms showed better receiver operating characteristic area under the curve (ROC-AUC) performance on lighter shades of skin. Despite efforts to improve DeepDerm's training methods, the performance gap between these subsets persisted.

Their research assessed the sensitivity of the algorithms' performance across skin tones using the DDI dataset to detect malignancies, as well as uncommon and benign lesions. Highlighted with their findings were listed *challenges* that arise when detecting cutaneous malignancies with AI algorithms:

1. Substantially worse performance of state-of-the-art AI algorithms on lesions on dark skin compared to light skin based on biopsy-proven malignancies

2. The drop-off in the overall performance of AI algorithms developed from previously described data when benchmarked on DDI

Commonly used visual consensus label performance from dermatologists for training AI models has differences across skin tones and uncommon conditions (Daneshjou et al., 2021):

– ModelDerm ROC-AUC on DDI dataset: 0.65 (95% confidence interval [CI] 0.61–0.70)

– DeepDerm ROC-AUC on DDI dataset: 0.56 (95% CI 0.51–0.61)

– HAM 10000 ROC-AUC on DDI dataset: 0.67 (95% CI 0.62–0.71)

– ModelDerm ROC-AUC on Fitzpatrick I to II subset: 0.64

– ModelDerm ROC-AUC on Fitzpatrick V to VI subset: 0.55

– DeepDerm ROC-AUC on Fitzpatrick I to II subset: 0.61

– DeepDerm ROC-AUC on Fitzpatrick V to VI subset: 0.50

– HAM 10000 ROC-AUC on Fitzpatrick I to II subset: 0.72

– HAM 10000 ROC-AUC on Fitzpatrick V to VI subset: 0.57

The results for retraining the algorithms on the DDI dataset through **transfer learning** showed decreased performance as compared to the results for their original test sets measured using the standard benchmarks. It was stated that performance limitations with these algorithms lie in the lack of diverse training data from the original experiment design, rather than the methods. Transfer learning has been used to improve medical imaging applications as well as cancer subtype discovery in genome sequencing (Hajiramezanali et al., 2018).

9.3 Defining Types of Biases and Mitigation Techniques in ML Life Cycles

Bias in ML can cause a wide array of adverse effects including discrimination. Bias in ML for health care results in algorithmic bias in applications that are intended to help people but end up reproducing bias. For example, Black patients are reported to have higher medical costs for emergency visits and lower outpatient specialist costs than white patients (Obermeyer et al., 2019). This signifies that Black patients are often forced to wait until it is an emergency to get care, whereas white patients have the privilege of getting routine nonemergency and preventative care. Bias in health care is a major barrier to access to care, resulting in African Americans' reported mistrust of the health care system, showing that race can directly affect health care (Armstrong et al., 2007; Obermeyer et al., 2019). This being the case, it is even more important to have ML for health care be available for Black patients to aid in triaging and diagnosing diseases such as skin cancer and working toward mitigating racial bias in general.

At each stage of development, bias risks must be evaluated and proactively addressed. Following is a summary of the biases covered in this chapter: **cognitive, evaluation, sampling, statistical, and underestimation** biases. Table 9.2 summarizes a brief definition, synonyms, position in the ML development life cycle, and mitigation methods for each bias, which are covered next. Think about how these biases are relevant in the case study of mitigating skin tone bias in ML for melanoma detection.

Cognitive biases are reflected in ML algorithms from humans, stemming from real-world inequity and discrimination, which can be propagated through data and result in predictions and decisions that are "unfair" (Angwin et al., 2022). *Unfairness* in ML decision making equates to prejudice or favoritism based on inherent or acquired characteristics of an individual or group (Mehrabi et al., 2021). Sometimes, unconscious prejudices from health care workers and researchers can introduce bias, which leads to systemically skewed data collection, including clinical trials which are predominantly carried out on Caucasian male patients.

Cognitive bias can occur at any stage of the ML development life cycle and has also been called social bias, historical bias, societal bias, individual bias, preexisting bias, negative legacy, or health care bias, as described in this chapter's case study. Possible mitigation strategies could be inclusion of diversity at all levels, as well as education around the importance of inclusion and diversity.

Statistical bias is a built-in or naturally occurring error that indicates the amount by which all observed values are wrong and can come from all stages of data analysis. Statistical bias in ML can be easily understood in relation to simple ANNs, which inspired the perceptron algorithm (McCulloch & Pitts, 1943; Rosenblatt, 1958). Perceptions are a form of supervised learning which uses training data to learn a link between inputs and outputs.

The perceptron function is a linear binary classifier, mapping one vector consisting of a real value as the input represented as x to an output value of 1 or 0 represented as $f(x)$. The 1 or 0 outputs classify x as a positive or negative instance of the function.

ANN Perceptron: $f(x) = \{1 \; if \; w * x + b > 0; 0 \; otherwise.$ (9.1)

The **bias** b in Eq. 9.1, also known as the bias of an estimator, should not be confused with the cognitive bias as described earlier. To calculate the estimator bias, you take the mean of estimated differences by adding up the errors in each estimate and compare them to the true values of the parameter being estimated, then divide by the number of estimates. The ultimate goal is to have a low-bias estimator with low variance in the predicted outcomes.

The bias added to the perceptron equation in Figure 9.8, is used to shift the decision boundary away from the starting point (origin), improving the accuracy of our calculations and predictions.

Statistical bias is found when the expected value of the results differs from the true value being estimated. Statistical bias occurs in all stages of the ML pipeline, including data selection, hypothesis testing, estimator selection, analysis, and interpretation. The bias of an estimator (or bias function) here represents the difference between estimated and actual values. Understanding this bias is useful because it tells us if our model is flexible enough to be able to calculate the average true relationship between weighted variables. Techniques for mitigating the error associated with statistical bias include in preprocessing and increasing the size of the dataset for training in order to reduce variability. Or, during training, you can incorporate a loss function, known as taking the mean square error (MSE), in order to reduce statistical bias and variance simultaneously.

Sampling bias occurs in preprocessing, where results from the dataset are nonrepresentative of the population of the intended application. For example, with only white skin images being utilized to train the algorithm that designed is for a broader, more diverse population (Adamson & Smith, 2018; Vokinger et al., 2021).

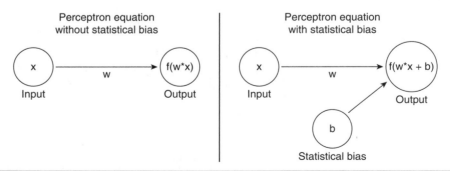

Figure 9.8   Perceptron equation with and without statistical bias.

Parallel ML Bias History: Similarly to the example of melanoma detection algorithms, in a study of facial recognition by Joy Buolamwini and Timnit Gebru, fewer cases of humans with darker faces were found in the ImageNet dataset, and an underlying *sampling bias* existed in the resulting models which were trained on it (Buolamwini & Gebru, 2018). When facial recognition was first developed, it lacked training on dark-skinned women and testing on benchmarks with similar non-inclusiveness. In order to mitigate sampling bias, one approach would be to create processes for tracking changes in social norms and the escalation of potential harm resulting from ML biases that would be established.

Evaluation bias can be introduced when a model is trained and evaluated on a dataset that uses inappropriate, vague, nondescriptive, and/or noninclusive performance metrics and thus is nonrepresentative of the population (Van Giffen et al., 2022) This can lead to overestimation and inaccuracies in the postprocessing stage. Some mitigation methods could be the use of inclusive datasets for training and testing the algorithm and relabeling data to match the truth.

Underestimation bias occurs when a model *underfits* the data and is not able to generalize to unseen data. Factors contributing to underestimation bias include limitations in the training data such as class imbalances and underrepresented categories, as well as model capacity issues such as *irreducible error* (Blanzeisky & Cunningham, 2021).

Regularization methods are used to reduce error in generalization by reducing model complexity to control for variance which leads to overfitting and underfitting of the algorithm. Model complexity can be reduced by limiting the number of nodes in the hidden layer or adding a penalty term in the loss function. Excessive use of regularization mechanisms in the case of an algorithm overfitting can overcorrect and result in underfitting or underestimation of predictions. Underestimation is influenced by excessive regularization mechanisms for bias and variance of a model.

Mitigation of underestimation bias can and should be addressed in all steps of the ML development life cycle. During preprocessing, the sample size of the minority group data can be increased to reduce class imbalances and underrepresented categories. During in-processing, cost-sensitive learning algorithms can be applied for imbalance classification. Cost-sensitive techniques may be divided into three groups, including data resampling, algorithm modifications, and ensemble methods. A specific example of algorithm modification includes considering underestimation in hyperparameter tuning. During postprocessing, underestimation bias may be mitigated by adjusting the optimal threshold value for the minority group. Table 9.2 describes the types of bias which were found and mitigated in the Stanford DDI research for skin cancer detection.

Biases in ML often occur due to limitations in the model and development process listed in Table 9.3. Understanding the limitations can help researchers and end users take steps to identify biases and mitigate them in each stage of the ML development life cycle.

Type of Bias	Example: Melanoma Case Study	Mitigation Methods in Melanoma Case Study
Evaluation bias	Algorithm trained and tested on light skin, does not represent the whole population	Use of datasets with diverse skin tones for training and testing the algorithm Retraining with the DDI dataset
Statistical bias	Higher false negatives for cancerous cells on dark skin tones Low accuracy of positive melanoma detection on dark skin	Increased size of the dataset for training Retraining with the DDI dataset Reduce statistical bias alongside variance in data by including a loss function (aka MSE)
Cognitive bias	Historical studies done on only white individuals for skin cancer; dermatology textbooks rarely show nonwhite skin	Inclusion of diversity at all levels; education around the importance of inclusion and diversity Education for health professionals and individuals regarding the prevalence of melanoma and implications for dark-skinned patients
Sampling bias	Previous datasets only represented light skin, omitting anyone with dark skin	Retrain algorithm on Diverse Dermatology Images (DDI) dataset

TABLE 9.2   Types of Bias in Melanoma Detection and Mitigation Methods

Limitation	Description	Correlated Bias(es)
Constrained time budget	Time and cost required to train machine learning models accurately can be high	Evaluation bias, sampling bias, statistical bias, labeling bias
Requires large datasets	Unable to learn from limited training examples	Sampling bias, statistical bias
Vanishing gradient	More layers than needed can lead to degradation of accuracy (saturation)	Statistical bias, underestimation bias
Not generalizable	Knowledge from one task can only be transferred to similar tasks	Underestimation bias
Lacks understanding	No commonsense knowledge of the world or the data it is being trained on	Cognitive bias
Lacks creativity or imagination	Not useful for tasks beyond classification or dimensionality reduction on their own	Cognitive bias

TABLE 9.3   Limitations of Machine Learning Models and Correlated Bias

9.4   Machine Learning Fairness

Fairness in ML can primarily be achieved through algorithmic fairness strategies and techniques. *Algorithmic fairness* is a mathematical attempt to obtain better outcomes in order to treat different groups equally.

Mitigation of bias in ML models often begins with understanding and implementing the previous methods when appropriate throughout the ML development life cycle including data collection, data processing, model training and evaluation, model deployment, and continuous improvement. This chapter describes strategies for mitigating bias at every stage of the ML development life cycle as depicted in Figure 9.9.

- *Preprocessing* bias mitigation techniques attempt to remove discrimination by adding more data or modifying the available training data.
- *In-processing* bias mitigation techniques affect the algorithm itself and the learning procedure by imposing constraints, updating the objective function, or regularization.
- *Postprocessing* bias mitigation techniques may be implemented following model deployment or during the re-evaluation period in which adjustments are made to the model decision thresholds or the model output, including relabeling.

Technical mitigation methods should be complemented with nontechnical measures, for example, combating both cognitive bias and bias in the data, such as statistical

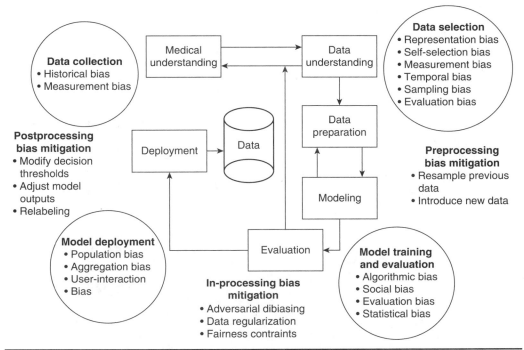

FIGURE 9.9   ML life cycle with bias indicators and mitigation techniques.

(inspired by Herhausen & Fahse, 2022; Huang et al., 2022; and van Giffen et al., 2022)

and evaluation bias. Before attempting to apply bias mitigation techniques, it is imperative to understand the sources and types of bias that may occur.

9.5 Chapter Summary

Whether biases in ML for health care are coming from systemic marginalization and social institutions (cognitive biases) or from flaws in the data or algorithm (statistical, estimation, sampling, and underestimation biases), sometimes they are easily detected and countered, but in many cases biases are hidden and untraceable (Starke et al., 2021). ML technology benefits those who are represented in the training data for a health care algorithm, leading to a loss of equity in treatment (sampling bias). Inequities are exemplified by the composition of training data for algorithms (Pot et al., 2021). Many health care professionals and researchers utilize the labels "black" and "white," terms that carry a particular cultural meaning, which are "not neutral, descriptive categories to classify people" and are "intrinsically entwined with the history of racism" (Pot et al., 2021). These terms influence how patients are treated.

As demonstrated through various research, including the DDI case study, bias in ML is a ubiquitous and challenging problem. Most bias is introduced unintentionally and can be very difficult to detect, and changes can trigger ML biases during operation or through retraining, feedback loops, or by the application context changing (Van Giffen et al., 2022). Whether or not to include race or skin tone in the metrics that are programmed into algorithms used to aid in health care is very situationally dependent. In the case of skin cancer, which presents differently on different skin tones, it is important that the algorithm is not "color blind."

Previous to the DDI dataset, all the major melanoma detection algorithms were flawed based on limited datasets. The perpetual bias of ML diagnostic systems would be harmful without the inclusion of diverse skin tones in the initial datasets. If bias is appropriately mitigated, diagnostic technology can be available to everyone and help save lives through accurate, early detection of skin disease. As we move forward using more advanced and involved ML systems in health care, it is vital to have better research, data, and practices which work toward inclusivity.

There are also researchers creating AI-generated images of skin lesions which not even dermatologists can recognize as different from real images (Figure 9.10). Is this a possible solution?

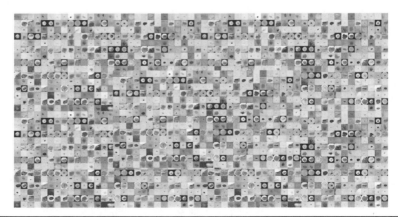

Figure 9.10 Computer-generated images of melanoma lesions.
(Kvak et al., 2023)

It is complex, cumbersome, and sometimes impossible to control for all types of bias in the dataset curation stage, but building models that incorporate algorithmic fairness is optimal considering the availability of data (Adeli et al., 2021). Bias-free ML does not exist, nor is there a panacea for all ML biases. Bias can result from many sources, including people, data, algorithms, or application processes, and can occur throughout the entire project.

Chapter Glossary

Term	Definition
Artificial neural network (ANN)	An algorithm that maps inputs from the dataset to outputs for problems like regression or classification, iteratively using a set of weights calculated at each node in the network and taking the summation of those weights through an optimization algorithm.
Conditional statistical parity	States that people in both protected and unprotected (e.g., female and male) groups should have equal probability of being assigned to a positive outcome given a set of legitimate factors.
Decision tree (DT)	A graphical representation of a decision-making process that uses a tree like model of decisions and their possible outcomes. (GeeksforGeeks, 2023)
Demographic parity	Also known as statistical parity, the likelihood of a positive outcome should be the same regardless of whether the person is in the protected (e.g., female) group.
Equal opportunity	The probability of a person in a positive class being assigned to a positive outcome should be equal for both protected and unprotected (e.g., female and male) group members (equal true positive rates).
Equalizing odds	Check if, for any particular label and attribute, a classifier predicts that label equally well for all values of that attribute. (Machine Learning Glossary, n.d.)
Fairness in relational domains	Captures the relational structure in a domain not only by taking attributes of individuals into consideration but also by taking into account the social, organizational, and other connections between individuals.
Fairness through awareness	An algorithm is fair if it gives similar predictions to similar individuals (based on distance on a graph).
Fairness through unawareness	An algorithm is fair as long as any protected attributes are not explicitly used in the decision-making process.
Irreducible error	Also known as Bayes error. The lower limit of the error that you can get with any classifier. A classifier that achieves this error rate is an optimal classifier. (Wehbe, 2019)
Overfitting	A concept in data science that occurs when a statistical model fits exactly against its training data. When this happens, the algorithm unfortunately cannot perform accurately against unseen data, defeating its purpose. (What is underfitting?, n.d.)
Supervised learning	A subcategory of machine learning and artificial intelligence that uses labeled datasets to train algorithms that classify data or predict outcomes accurately. As input data is fed into the model, it adjusts its weights until the model has been fitted appropriately. (What is supervised learning?, n.d.)

(Continued)

Term	Definition
Treatment equality	When the ratio of false negatives and false positives is the same for both protected group categories.
Underfitting	A scenario in data science where a data model is unable to capture the relationship between the input and output variables accurately, generating a high error rate on both the training set and unseen data. It occurs when a model is too simple, which can be a result of a model needing more training time, more input features, or less regularization. (What is underfitting?, n.d.)
Unsupervised learning	Uses machine learning algorithms to analyze and cluster unlabeled datasets. These algorithms discover hidden patterns or data groupings without the need for human intervention.

References

Adamson, A. S., & Smith, A. (2018). Machine learning and health care disparities in dermatology. JAMA Dermatology, 154(11), 1247. https://doi.org/10.1001/jamadermatol.2018.2348

Adeli, E., Zhao, Q., Pfefferbaum, A., Sullivan, E. V., Fei-Fei L., Niebles, J. C., & Pohl, K. M. (2021). Representation learning with statistical independence to mitigate bias. 2021 IEEE Winter Conference on Applications of Computer Vision (WACV). https://doi.org/10.1109/wacv48630.2021.00256

American Academy of Dermatology. (2016, July). Poor skin cancer survival in patients with skin of color, study shows. Science Daily. www.sciencedaily.com/releases/2016/07/160728102204.htm

Angwin, J., Larson, J., Mattu, S. & Kirchner, L. (2022). "Machine bias." In Ethics of data and analytics (pp. 254–264). Auerbach Publications.

Armstrong, K., Ravenell, K. L., McMurphy, S., & Putt, M. (2007). Racial/ethnic differences in physician distrust in the United States. American Journal of Public Health, 97(7), 1283–1289. https://doi.org/10.2105/AJPH.2005.080762

Blanzeisky, W., & Cunningham, P. (2021). Algorithmic factors influencing bias in machine learning. Communications in Computer and Information Science, first published online, 559–574. https://doi.org/10.1007/978-3-030-93736-2_41

Buolamwini, J., & Gebru, T. (2018). Gender shades: Intersectional accuracy disparities in commercial gender classification. Proceedings of the 1st Conference on Fairness, Accountability and Transparency, PMLR, 81, 77–91.

Daghrir, J., Tling, L., Bouchouicha, M., & Sayadi, M. (2020). Melanoma skin cancer detection using deep learning and classical machine learning techniques: A hybrid approach. 2020 5th International Conference on Advanced Technologies for Signal and Image Processing (ATSIP). https://doi.org/10.1109/atsip49331.2020.9231544

Daneshjou, R., Vodrahalli, K., Liang, W., Novoa, R. A., Jenkins, M., Rotemberg, V., Ko, J., et al. (2021). Disparities in dermatology AI: Assessments using diverse clinical images.

DDI Alliance. (2022). DDI – Data Documentation Initiative. https://ddi-dataset.github.io/index.html#dataset

Dhivyaa, C. R., Sangeetha, K., Balamurugan, M., Amaran, S., Vetriselvi, T., & Johnpaul, P. (2020). Skin lesion classification using decision trees and random forest algorithms. Journal of Ambient Intelligence and Humanized Computing. https://doi.org/10.1007/s12652-020-02675-8

Esteva, A., Kuprel, B., Novoa, R. A., Ko, J., Swetter, S. M., Blau, H. M., & Thrun, S. (2017). Dermatologist-level classification of skin cancer with deep neural networks. Nature, 542(7639), 115–118. https://doi.org/10.1038/nature21056

Fitzpatrick, T. B. (1975). Soleil et peau. Journal of Esthetic Dentistry, 2(January), 33–34.

Google. (2023). Machine Learning Glossary: Fairness | Google Developers. https://developers.google.com/machine-learning/glossary/fairness#fairness_metric

GeeksforGeeks. (2023, March 6). Decision Tree. https://www.geeksforgeeks.org/decision-tree/

Hajiramezanali, E., Dadaneh, S. Z., Karbalayghareh, A., Zhou, M. & Qian, X. (2018). Bayesian multi-domain learning for cancer subtype discovery from next-generation sequencing count data. Advances in Neural Information Processing Systems, 31.

Han, S. S., Park, I., Chang, S. E., Lim, W., Kim, M. S., Park, G. H., Chae, J. B., Huh, C. H., Na, J.-I. (2020). Augmented intelligence dermatology: deep neural networks empower medical professionals in diagnosing skin cancer and predicting treatment options for 134 skin disorders. Journal of Investigative Dermatology, 140(9), 1753–1761. https://doi.org/10.1016/j.jid.2020.01.019

Hekler, A., Utikal, J. S., Enk, A. H., Hauschild, A., Weichenthal, M., Maron, R. C., Berking, C., et al. (2019). Superior skin cancer classification by the combination of human and artificial intelligence. European Journal of Cancer, 120(October), 114–121. https://doi.org/10.1016/j.ejca.2019.07.019

Hu, S., Soza-Vento, R. M., Parker, D. F., & Kirsner, R. S. (2006). Comparison of stage at diagnosis of melanoma among Hispanic, Black, and White patients in Miami-Dade County, Florida. Archives of Dermatology, 142(6), 704–708. https://doi.org/10.1001/archderm.142.6.704

Huang, J., Galal, G., Etemadi, M., & Vaidyanathan, M. (2022). Evaluation and mitigation of racial bias in clinical machine learning models: Scoping review. JMIR Medical Informatics, 10(5). https://doi.org/10.2196/36388

IBM. (2023a). What Is Supervised Learning? https://www.ibm.com/topics/supervised-learning?mhsrc=ibmsearch_a&mhq=Supervised+Learning

IBM. (2023b). What Is Underfitting? https://www.ibm.com/topics/underfitting#:~:text=the%20next%20step-,What%20is%20underfitting%3F,training%20set%20and%20unseen%20data

Jaworek-Korjakowska, J. (2016). Computer-aided diagnosis of micro-malignant melanoma lesions applying support vector machines. BioMed Research International, 2016: 1–8. https://doi.org/10.1155/2016/4381972

Kato, J., Horimoto, K., Sato, S., Minowa, T., & Uhara, H. (2019). Dermoscopy of melanoma and non-melanoma skin cancers. Frontiers in Medicine, 6(August). https://doi.org/10.3389/fmed.2019.00180

Kaur, R., Gholamhosseini, H., Sinha, R., & Lindén, M. (2022). Melanoma classification using a novel deep convolutional neural network with dermoscopic images. Sensors, 22(3). https://doi.org/10.3390/s22031134

Kvak, D., Biroš, M., Hrubý, R., & Březinová, E. (2023). Synthetic data as a tool to combat racial bias in medical AI: Utilizing generative models for optimizing early detection of melanoma in Fitzpatrick skin types IV-VI. In Proceedings of 2022 International Conference on Medical Imaging and Computer-Aided Diagnosis (MICAD 2022) Medical Imaging and Computer-Aided Diagnosis. Springer Singapore.

McCulloch, W. S., & Pitts, W. (1943). A logical calculus of the ideas immanent in nervous activity. The Bulletin of Mathematical Biophysics, 5(4), 115–133. https://doi.org/10.1007/BF02478259

Mehrabi, N., Morstatter, F., Saxena, N., Lerman, K., & Galstyan, A. (2021). A survey on bias and fairness in machine learning. ACM Computing Surveys, 54(6), 1–35. https://doi.org/10.1145/3457607

Mendonca, T., Ferreira, P. M., Marques, J. S., Marcal, A. R. S., & Rozeira, J. (2013). PH2 – A dermoscopic image database for research and benchmarking. In 2013 35th Annual International Conference of the IEEE Engineering in Medicine and Biology Society (EMBC), 5437–5440. IEEE. https://doi.org/10.1109/EMBC.2013.6610779

Micu, E. (2014). Solar radiation and skin cancer risk-biophysical insight. Romanian Journal of Biophysics, 24(4).

Murugan, A., Nair, S. A. H., & Kumar, K. P. S. (2019). Detection of skin cancer using SVM, random forest and KNN classifiers. Journal of Medical Systems, 43(8), 269. https://doi.org/10.1007/s10916-019-1400-8

Obermeyer, Z., Powers, B., Vogeli, C., & Mullainathan, S. (2019). Dissecting Racial Bias in an Algorithm Used to Manage the Health of Populations. http://science.sciencemag.org/

Pot, M., Kieusseyan, N., & Prainsack, B. (2021). Not all biases are bad: Equitable and inequitable biases in machine learning and radiology. Insights into Imaging, 2. https://doi.org/10.1186/s13244-020-00955-7

Rosenblatt, F. (1958). The perceptron: A probabilistic model for information storage and organization in the brain. Psychological Review, 65(6), 386–408. https://doi.org/10.1037/h0042519

Rotemberg, V., Kurtansky, N., Betz-Stablein, B., Caffery, L., Chousakos, E., Codella, N., Combalia, M., et al. (2021). A patient-centric dataset of images and metadata for identifying melanomas using clinical context. Scientific Data, 8(1), 34.

Starke, G., de Clercq, E., & Elger, B. S. (2021). Towards a pragmatist dealing with algorithmic bias in medical machine learning. Medicine, Health Care and Philosophy, 24(3). https://doi.org/10.1007/s11019-021-10008-5

Stegenga, J. (2018). Care and cure: An introduction to philosophy of medicine. In Care and Cure: An Introduction to Philosophy of Medicine. Chicago: The University of Chicago Press.

Suresh, H., Gong, J. J., & Guttag, J. V. (2018). Learning tasks for multitask learning. In Proceedings of the 24th ACM SIGKDD International Conference on Knowledge Discovery & Data Mining (pp. 802–810). New York, ACM. https://doi.org/10.1145/3219819.3219930

Thomsen, K., Iversen, L., Titlestad, T. L., & Winther, O. (2020). Systematic review of machine learning for diagnosis and prognosis in dermatology. Journal of Dermatological Treatment, 31(5), 496–510. https://doi.org/10.1080/09546634.2019.1682500

Tschandl, P., Rosendahl, C., & Kittler, H. (2018). The HAM10000 dataset, a large collection of multi-source dermatoscopic images of common pigmented skin lesions. Scientific Data, 5(1), 180161. https://doi.org/10.1038/sdata.2018.161

Van Giffen, B., Herhausen, D., & Fahse, T. (2022). Overcoming the pitfalls and perils of algorithms: A classification of machine learning biases and mitigation methods. Journal of Business Research, 144, 93–106. https://doi.org/10.1016/j.jbusres.2022.01.076

Vokinger, K. N., Feuerriegel, S., & Kesselheim, A. S. (2021). Mitigating bias in machine learning for medicine. Communications Medicine, 1(1). https://doi.org/10.1038/s43856-021-00028-w

Wehbe, L. (2019). The School of Computer Science at Carnegie Mellon Course 10-701 Lecture 3. https://www.cs.cmu.edu/~lwehbe/10701_S19/files/Lecture_3.pdf

World Cancer Research Fund. (2022). Skin Cancer Statistics. https://www.wcrf.org/dietandcancer/skin-cancerstatistics/

End of Chapter Questions

Match questions 1 to 5 with answers a to e.

1. What is sampling bias?

2. What is evaluation bias?

3. What is statistical bias?

4. What is cognitive bias?

5. What is underestimation bias?

   a. When a model is trained and evaluated using a dataset that is nonrepresentative

   b. Results from the dataset are nonrepresentative of the population of intended use

   c. A built-in or naturally occurring error that indicates the amount by which all observed values are wrong

   d. Human bias that can result in predictions and decisions that are "unfair"

   e. When a model underfits the data and is not able to generalize to unseen data

6. How do algorithms differ from humans in decision making?

7. What are the recommended actions for mitigating bias in ML? (Select all correct answers.)

   a. Bias risks must be evaluated and proactively addressed at all stages

   b. Must document assumptions and decisions regarding ML applications

   c. Processes to discover bias proactively must be established during development and implementation

   d. Establish processes for tracking changes in social norms and the escalation of potential harm resulting from ML biases

   e. Include end users in the co-development and prototyping of ML applications for transparency

   f. All of the above

8. True or False: People with melanated skin can't get skin cancer.

9. What are the ABCDE procedure characteristics for identifying melanoma?

10. What ML processes are used in melanoma detection?

11. Why is it important to train AI algorithms for melanoma detection on a diverse skin tone dataset?

CHAPTER 10

Applying the Wells-DuBois Protocol for Achieving Systemic Equity in Socioecological Systems

Ayushi Aggarwal

Civil & Environmental Engineering, Georgia Institute of Technology, Atlanta

Tyrek Shepard

School of Public Policy, Georgia Institute of Technology, Atlanta

Thema Monroe-White

Department of Technology, Entrepreneurship and Data Analytics, Berry College, Mount Berry

Joe F. Bozeman III

Civil & Environmental Engineering, Georgia Institute of Technology, Atlanta
School of Public Policy, Georgia Institute of Technology, Atlanta

Question: What overarching tools and concepts should you use when attempting to yield equitable outcomes in social and ecological (socioecological) systems?

Learning Objectives

Upon completion of this chapter, the student should be able to

- Understand how the Wells-DuBois Protocol helps to achieve systemic equity
- Apply a machine learning (ML) approach to a socioecological dataset

- Evaluate the bias or inequity of artificial intelligence (AI) and ML datasets in socioecological systems using the Wells-DuBois protocol
- Remember that there is a need to further refine systemic equity approaches in AI/ML applications

Chapter Overview

This chapter begins by introducing a key equity framework (i.e., the systemic equity framework) and tool (i.e., the Wells-DuBois protocol) for mitigating bias or avoiding inequities in AI and ML activities. It then provides explanation and coding details for an ML-clustering application case that involves food systems and race/ethnicity as the socioecological context. The application of the Wells-DuBois protocol follows with examples for how to answer its questions. The chapter then moves to a conclusion with a discussion on pertinent equity-centered challenges and future directions. Lastly, several chapter problems are provided to assist with concept retention. The overall learning objective of this chapter is to effectively evaluate the bias or inequity of AI and ML datasets in socioecological systems.

10.1   Introduction

There are many entry points into AI and ML. Nonetheless, this section provides introductory material on some key AI and ML applications that interact with social equity. The primary learning objective of this section is to gain a general understanding for how AI and ML can yield bias or inequity in social and ecological (i.e., socioecological) systems.

10.1.1   Understanding AI and ML Use in Socioecological Systems

Recent decades have seen the integration of AI and ML into various dimensions of modern socioecological systems, commerce, and countless sectors that leverage advanced technologies (Akter et al., 2022; Özyüksel Çiftçioğlu & Naser, 2022). AI refers to the intelligence demonstrated by computational algorithms that mimic human-like intelligence. ML, on the other hand, is a subset of AI that uses continuous learning methodologies to solve complex problems.

The surfacing of sophisticated data techniques has unlocked new potential for innovation and novel research applications (Borana, 2016; Das et al., 2015; Nadimpalli, 2017), but those leveraging these techniques must acknowledge the biases and flawed outcomes that can be produced when using data-centered tools and models. Bias—or inequity, which will be used interchangeably throughout this chapter—generally refers to unfair outcomes in AI and ML applications across factors such as problem formation, sociodemographic subgroups (e.g., race, ethnicity, and socioeconomic status), model design and outcomes, and interpretation. There are more details on bias and socioecological inequity in Section 1.2.

Accurate forecasting, modeling, and artificial replications of human-like systems that synthesize data better than humans are only a sample of the possibilities with AI/ML (Weerasuriya et al., 2021). Despite the benefits, roadblocks exist in unlocking equitable and human-centered AI tools (Akter et al., 2022). These tools have been found to perpetuate the biases of their training sets since they are trained by human-selected data (Balayn et al., 2021; Leavy, 2018; Tae et al., 2019), yielding biased outcomes that impact our socioecological systems (Donati et al., 2022; Galaz et al., 2021; Green, 2018). These computationally driven inequities have tangible impacts on society.

10.1.2   Examples of Socioecological Inequity and Bias

One example of how data-driven techniques can yield inequity in socioecological systems is at the nexus of health and nutrition. Bias can manifest in the dietary components and energy intake reporting of the National Health and Nutrition Examination Survey (NHANES) (Moshfegh et al., 2008; Steinfeldt et al., 2013). The U.S. Department of Agriculture's Automated Multiple Pass Method is the data collection technique that informs the NHANES; however, it remains unclear how the information pertaining to a broader spectrum of race and demographic factors was collected (Pannucci et al., 2018). It follows that human behavior within the food system varies across sociodemographic subgroups (Bozeman et al., 2020), making it necessary to integrate sociodemographic considerations when attempting to be fair and equitable.

There are several other socioecological examples in this regard. For instance, digital twins are virtual proxies of a real and physical system that is updated in real time, powered by ML, and is helpful for tracking and decision making (Onile et al., 2021). This technology has the potential for use in energy demand management to ensure sustainable energy generation and distribution for consumers (Huang et al., 2022). Although this is a novel way of ensuring consumer-driven utility delivery, data collected about energy use may exclude customers' sociodemographic attributes (de Ayala et al., 2020; Bozeman, Chopra et al., 2022).

AI tools have the potential to serve users across a variety of other products and disciplines but run the risk of inequitable outcomes if mismanaged. Facial recognition is one of the more common and controversial uses of AI technology (Andrejevic & Selwyn, 2020; Brey, 2004; Buolamwini & Gebru, 2018). Several products intended for photograph classification and organizing have delivered distressing results to users within recent years, such as Google Photos' 2015 incident where African American users were categorized as "gorillas," along with reports that the same application was classifying white users as dogs and seals ("Google apologises for Photos app's racist blunder," 2015). These incidents exemplify insensitive and inequitable outcomes due to a lack of sociodemographic consideration in facial recognition modeling activities (Daugherty et al., 2019; Murray et al., 2020; Turner Lee, 2018).

10.1.3   Clustering in ML for Model Outcome Assessment

Clustering is an unsupervised ML application that groups data into different classes or categories. These categories can be predefined (i.e., before model deployment) or later defined as data associations reveal themselves. Put more simply, clustering groups data elements as the machine "learns" about the similarities and differences of these data. This feature allows for the detection of patterns that might otherwise be missed.

Clustering, as a visualization tool, has been used to assess bias and equity in model outcomes. Some examples include the forecasting of education levels in Twitter users (Florea & Roman, 2021) and the analysis of cultural bias in an opinion-based survey on university students' individual learning experiences (Owsiński et al., 2022). ML applications are inherently flawed if they perpetuate societal inequities or create isolating experiences. An accumulation of these occurrences has led to research on how to mitigate or avoid these inequities. The tools and concepts discussed in the subsequent chapter sections are inspired by these types of findings.

10.1.4   Basic Concepts and Definitions

Socioecological system: A system that involves both social and ecological components.

   Systemic equity: A comprehensive framework that helps to mitigate inequitable outcomes and can only be achieved when distributive, procedural, and recognitional equities are addressed simultaneously and over an extended period.

   Wells-DuBois protocol: A checklist and tool made up of seven components and three categories that helps to achieve systemic equity by identifying and assessing biases in AI/ML applications.

10.2   Equity Framework and Tool Application

In this section, we overview a key equity framework and tool that, when administered together, can help mitigate or minimize inequities in the application of AI/ML approaches. Furthermore, this framework and tool are applied to an ML-clustering scenario. The learning objectives are to remember the three core components of systemic equity, understand how the Wells-DuBois protocol can be used as a tool to help achieve systemic equity, and learn how to apply the Wells DuBois protocol in an ML context.

10.2.1   The Systemic Equity Framework

The integration of equity-centered frameworks and tools can help technologists and decision makers identify bias within data and model outcomes. Identification is the first step toward mitigating or preventing inequitable outcomes in AI/ML applications. *Systemic equity*, a comprehensive framework that helps to mitigate inequitable outcomes (see Figure 10.1), can only be achieved when distributive (i.e., the act of providing tangible resources to a person or group in an unbiased and fair manner), procedural (i.e., the act of employing decision-making activities that facilitate the allocation of resources in an unbiased and fair manner), and recognitional equities (i.e., addressing

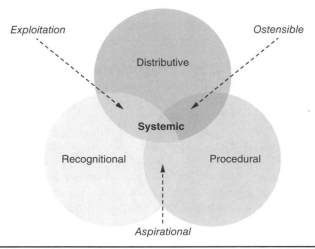

FIGURE 10.1   Schematic of the systemic equity framework.

the psychological, emotional, and cultural needs of the systematically marginalized where bias and disadvantage are embedded or evident) are addressed simultaneously and over an extended period (Bozeman, Nobler et al., 2022; Howard, 2016).

Ineffective equities or the two application unions (i.e., exploitational, ostensible, and aspirational equities) of the systemic equity framework are now explained. As observed in Figure 10.1, ostensible equity presents itself when only distributive and procedural equities are met but recognitional equity is not. Aspirational equity occurs when procedural and recognitional equities are met but distributive equity is not. Lastly, exploitational equity—the third ineffective-equity categorization of this framework—manifests when both distributive and recognitional equities are met but procedural equity is not. In practice, systemic equity solutions may fall into one of these ineffective-equity categories at various junctures.

10.2.2   The Wells-DuBois Protocol

Technologists, decision makers, and the like must take a proactive and comprehensive approach in an effort to achieve systemic equity. The Wells-DuBois protocol is an apt tool and checklist for doing this (Monroe-White & Lecy, 2022). Named after two pioneering historical social activists and race-conscious data scholars, W.E.B. DuBois and Ida B. Wells (Monroe-White, 2021, 2022), it assesses if sociodemographic biases are qualified for when data science methodologies are employed. The Wells-DuBois protocol has seven components that are grouped into three categories (see Table 10.1).

The application of the Wells-DuBois protocol is simple. To apply it, answer the questions in Table 10.1 before incorporating data into any AI/ML model and adjust modeling activities accordingly. Following this protocol could prevent the need for tedious and time-consuming adjustments post model deployment. Furthermore, this protocol can be used to examine the equitability of existing applications.

10.2.3   Similarities and Differences with Other Equity Tools

The Wells-DuBois protocol and systemic equity framework share similarities with other initiatives, such as the data equity framework and Washington State's Pro-Equity Anti-Racism (PEAR) Plan & Playbook, in their commitment to addressing systemic biases with an equity-centered approach (Krause, 2017; PEAR, 2022). However, there are differences between these frameworks. While the data equity framework and PEAR Plan & Playbook prioritize data equity, social determinants of equity, and anti-racism efforts more broadly, the Wells-DuBois protocol and systemic equity framework take a more comprehensive approach, providing a detailed methodology for achieving systemic equity across multiple domains, including data, models, and decision-making processes.

A limitation of the data equity framework and PEAR Plan & Playbook is their lack of explicit guidance on evaluating and addressing bias in decision-making processes beyond the data itself. In contrast, the Wells-DuBois protocol and systemic equity framework address this gap by providing a more comprehensive approach that includes procedural and recognitional equities. Furthermore, while all four frameworks and tools prioritize some aspect of subpopulation inclusion, the Wells-Dubois protocol places a stronger emphasis on identifying subpopulation harm across all levels of the data use process, such as collection, algorithmic creation, and output review, which distinguishes it from the others.

Bad Data		
Inadequate Data		
Do the data overlook, erroneously represent, or systemically exclude a subpopulation?		
Tendentious Data		
Do the data represent the subjectivity or impartiality of humans? How does this bias affect the intended outcomes?		
Algorithmic Bias		
Harms of Identity Proxy		
Could the model treat a particular demographic differently, even without explicit identity markers?		
Harms of Subpopulation Difference		
Are algorithmic outcomes disparate across respective subgroups?		
Harms of Misfit Models		
If the models are predictive, have you examined their accuracy by subpopulation to ensure performance is not significantly different? Specifically, what is your value orientation and what are the public/social implications of this work?		
Human Intent		
Do No Harm		
What are your goals and intended outcomes? Is any ill intent involved?		
Harms of Ignorance		
What are the unintended consequences of your work? How can your results be manipulated to abuse or harm?		

TABLE 10.1   The Wells Du-Bois Protocol

To move toward systemic equity in AI/ML applications, one needs to first acknowledge, identify, and then try to mitigate biases present in the socioecological datasets of associated models. This can be done by employing the Wells-DuBois protocol. In the following sections, we overview the ML method of clustering and show how the Wells-DuBois protocol should be applied.

10.2.4 Section Summary

In this section, a key equity framework (i.e., the systemic equity framework) and tool (i.e., the Wells-DuBois protocol) that can help mitigate bias and inequities in the application of AI/ML approaches were reviewed. Then, it explained some of the similarities and differences with other equity tools (i.e., the data equity framework and Washington State's PEAR Plan & Playbook). As for learning objectives, this section provides understanding for how the Wells-DuBois protocol helps to achieve systemic equity.

10.3   Clustering Overview and Application

ML tools (e.g., clustering, neural network modeling, sentiment analysis, and natural language processing) involve complexities ranging from challenges in syntax and coding, to understanding the nuances of algorithmic purpose, and data visualization. Clustering, one of these ML tools, can be used for model outcome analysis, as previously mentioned. Clustering is therefore used given its relative ease in data visualization and concept comprehension. Furthermore, this relative ease in visualization is effective in identifying implicit and overt bias.

Now, to overview how clustering applications work. The first step in finding a relationship between data points is calculating the correlation between the *n* number of variables in a dataset. Correlation is the statistical estimation of the relational proximity between variables. This estimation can be positive or negative. A positive correlation means that as one variable changes in a direction, the other variable positively correlated to it will also change in the same direction. A negative correlation implies the opposite: as one variable changes in a direction, the other variable negatively correlated to it will change in the opposite direction. This produces a quantitative correlation between two or *n* number of variables.

Clustering also allows for a qualitative analysis of a dataset by visually plotting data points. This is practically performed by using, for example, the *scatter()* function in MATLAB or the *plt.scatter()* function in Python. Through these scatter plots, visible clusters or groups can be evaluated directly or undergo further analysis.

In situations where groupings are not distinct enough, clustering algorithms such as *k-means, hierarchical,* or density-based spatial clustering of applications with noise—also known as *DBSCAN*—can be applied (Tyagi, 2021). The basic mechanism for how each algorithm creates clusters differs and can vary in terms of their performance (Rodriguez et al., 2019). When the dataset is too complex or noisy for efficient clustering, it is customary to apply principal component analysis (PCA). PCA dimensionally reduces large and noisy datasets while preserving important information (Jaadi, 2021).

It is important to note that PCA use, in clustering and other ML applications, has been a subject of debate for many years. PCA effectively diminishes certain aspects of a dataset to emphasize others, potentially creating linear combinations (Everitt et al., 2011). For example, PCA use in facial image clustering can produce varying results on the same image due to differentiating interpretation of information in the presence of noisy data (Lam & Choy, 2019). On the other hand, PCA has been shown to be useful in removing noisy data to reveal outcomes that would otherwise be less apparent or hidden. For instance, PCA and clustering have been effective investigation tools in socioecological contexts such as food systems and human gene grouping (Ben-Hur & Guyon, 2003; Li et al., 2019; Miller et al., 2020). Further discussion on these matters is outside the scope of this chapter; however, understanding the implications of PCA use is key for more advanced applications.

10.3.1 A Socioecological Example on Food Spending

Data from the Consumer Expenditure (CE) survey of the U.S. Bureau of Labor Statistics (USBLS) serves as an ideal socioecological dataset (USBLS, 2022). This survey's data provide information on spending—or expenditures, income, and other demographic factors for consumers in the United States. Among other features, it breaks down the amount of U.S. dollars spent on major food items by socioeconomic status, race, education level, and gender. Looking at the 2021 income survey tabulation, there were 133,595 customer interviews and diary entries (i.e., individually written records) (USBLS, 2022).

As explained earlier, using clustering for analysis on such a demographically diverse dataset could yield biases or inequitable outcomes, whether wittingly or not. This is certainly the case for socioecological systems that involve factors such as economic and food-energy-water implications (Bozeman et al., 2019; Bozeman et al.,

2020). The main purpose of this subsection is to visualize how bias might emerge through clustering this CE dataset using MATLAB.

10.3.2 Visualization and Initial Analysis

As with most large datasets, it is important to perform an initial analysis before proceeding with more involved clustering applications. We did so by applying a basic correlation function on the 335 columns and approximately 2,965 rows of CE survey data that we attained from USBLS (2022). As anticipated, given the size of this dataset, there was no statistically viable correlation between columns. This indicated that analysis beyond the initial correlation should be performed. Specifically, we moved to an initial qualitative analysis via scatter plot visualization using MATLAB.

To assist our qualitative effort of scatter plot visualization, we manually selected columns that would be meaningful for this socioecological analysis by using the findings from (Bozeman et al., 2019). This previous study on U.S. food consumption impact rates across racial/ethnic (i.e., Black, Latinx, and white) and socioeconomic subgroups found that the average household income was correlated to racial/ethnic subgroups. It found that households that spend a higher proportion of their annual income on food (i.e., Black and Latinx households comparatively) are more likely to purchase cheaper, energy-dense, and less environmentally friendly food products. It follows that we selected food expenditures and annual individual income as the focal data point categories. After manually selecting these two categories with individual customer IDs, racial identity was selected as a scatter plot grouping factor (see Figure 10.2).

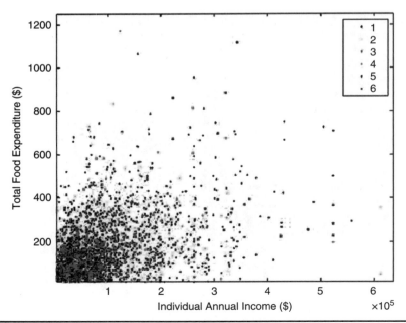

Figure 10.2 MATLAB scattering of food expenditures vs. individual annual income for a select group of interviewers. Here, 1–6 represents the different racial groups (i.e., 1 – White, 2 – Black, 3 – Native American, 4 – Asian, 5 – Pacific Islander, and 6 – Other races).

Figure 10.2 illustrates the first qualitative scatter plot for analysis. There are three observations worth highlighting: (1) most cluster points are of race 1 – Whites, (2) no clear groups are formed, and (3) most cluster points concentrate at the x-y-axis point—the lower-left corner of the plot. This initial scatter plot is important to establish before progressing toward more involved clustering applications.

Prior to employing PCA for more effective cluster administration, we selected individual annual income, education level, and total food expenditures as focal points per the guidance of Bozeman et al. (2019). As was the case in the initial scatter plot effort, race was chosen as the grouping factor for the clusters. Next, PCA was applied. We visualized the cluster of this refined dataset without race as a factor (see Figure 10.3a) and with race as an additional PCA factor (see Figure 10.3b) to preliminarily assess for bias in this regard. In terms of generic syntax, the data was represented as shown:

- PCA without race column

  X = [annual_income, education_lvl, tot_food_exp]

- PCA with race column

  X = [annual_income, education_lvl, tot_food_exp, race_ref]

Looking at Figure 10.3, one can observe that there are no clear differences between these two cluster visualizations. This shows that clustering scatter plots, in general, does not necessarily produce apparent or obvious indications of bias or inequity. However, findings from Bozeman et al. (2019) suggest otherwise.

To explore this dataset and its embedded biases further, we used an inbuilt clustering methodology in MATLAB. For clarity, a sample code is presented next for potential reproduction. Please note that guidelines to reproduce similar code in Python are provided after this sample code. The maximum number of clusters was set

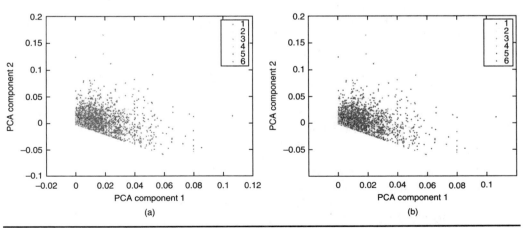

FIGURE 10.3   MATLAB-PCA scattering (a) without race and (b) with race, where racial groups 1 through 6 match that of Figure 10.2.

to six here to match the six racial groupings. It is also worth stating that, even when applying the exact same code, it is important to ensure ML applications have all the required functionalities (e.g., software updates, computer visualization capabilities, and the like).

Sample MATLAB code:

X1=[X(1:2952,1) X(1:2952,2) X(1:2952,4)]	%[annual_income, education_lvl, food_exp]
pca_comp=pca(X1')	%conducting PCA on selected columns
y=categorical(X(1:2952,3))	%refers to race_ref; categorically assign labels (race)
gscatter(pca_comp(:,1),pca_comp(:,2),y)	%plot scatter of PCAs and labels
T = clusterdata(pca_comp,'Linkage', 'ward','SaveMemory','on','Maxclust',6);	%clustering tool
scatter(pca_comp(:,1),pca_comp(:,2),10,T)	%plot clusters made using clustering tool
	%now conducting similar analysis but with addition of race column
pca_comp=pca(X(1:2952,1:4))')	%performing PCA over all 4 columns
gscatter(pca_comp(:,1),pca_comp(:,2),y)	%plot scatter of PCAs and labels
T = clusterdata(pca_comp,'Linkage', 'ward','SaveMemory','on','Maxclust',6);	%clustering tool
scatter(pca_comp(:,1),pca_comp(:,2),10,T)	%plot clusters made using clustering tool

A similar code in Python would use libraries like *numpy* and *matplotlib.pyplot*. In Python, inbuilt PCA syntax and tools like *plt.scatter(pca_comp[:,0], pca_comp[:,1], c=y)*, *PCA(n_components=2).fit_transform(X1.T)*, *AgglomerativeClustering(n_clusters=6, linkage= 'ward')*, and *cluster.fit_predict(pca_comp)* could be used to produce similar results.

It is evident that there are differences between Figure 10.4a and 10.4b, unlike in Figure 10.3's case. This suggests that the addition of race as a factor in the MATLAB-PCA clustering administration does change clustering outcomes compared to the plots of MATLAB-PCA scattering (refer to Figure 10.3). There are two observations worth emphasizing here: (1) the overall cluster sizes have changed across racial groups 1 through 6 when comparing Figure 10.4a to Figure 10.4b and (2) the planar (i.e., x-y-axis) location of the clusters has shifted (e.g., view the difference in location of the yellow cluster in Figure 10.4a and Figure 10.4b).

These three visualization approaches (i.e., MATLAB scattering, MATLAB-PCA scattering, and MATLAB-PCA clustering) produced varying results when presented with the same foundational dataset and problem. This suggests that more than simply applying ML methodology is needed when sociodemographic factors are present. Moreover, it exemplifies why equity-centered frameworks and tools are so important to employ in socioecological settings whether biases are apparent or not. These factors are vital when developing governmental policies or behavioral interventions that rely on datasets which embed sociodemographic information. An example of the

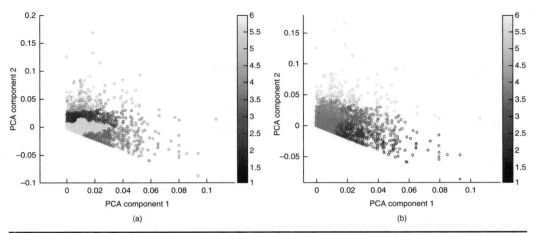

FIGURE 10.4 MATLAB-PCA clustering (a) without race and (b) with race, where racial groups 1 through 6 (i.e., the color-coded numerals located on the right side of each plot) match that of Figures 10.2 and 10.3.

same is discussed in the next section while exploring the application of the Wells-DuBois protocol.

10.3.3   Section Summary

In this section, the ML clustering approach was reviewed. Then, it explained how to apply this approach by providing explicit code in the MATLAB and Python languages. Illustrations were provided to show how the dataset application was visualized. This section provides guidance on how to apply an ML approach to a socioecological dataset.

10.4   Applying the Wells-DuBois Protocol

To explore an application of the Wells-DuBois protocol, let's consider our clustering example as the foundation for a policy designed to subsidize nutritious and sustainable foods. Subsidies are a common public policy tool used to incentivize or encourage certain actions. For example, providing financial assistance to qualifying households or directly discounting healthy inventory for stores in low-income neighborhoods can serve as a subsidy for nutritious and sustainable foods. Given further model analysis and forecast application, the results of our clustering example can serve as guidance for how to effectively initiate such a policy by, for example, identifying target populations that may benefit most.

The CE survey provides national-level data regarding the spending habits of U.S. citizens, along with several sociodemographic characteristics. As previously mentioned, Black and Latinx households are shown to spend a larger proportion of their respective incomes on cheaper, less sustainable foods. Even if the former insight were unknown, we can still leverage the Wells Du-Bois protocol to identify relevant points of concern. We apply it by answering the questions from Table 10.1 given our CE dataset, clustering application, and policy proposal (see Table 10.2).

Bad Data
Inadequate Data
Do the data overlook, erroneously represent, or systemically exclude a subpopulation?
Although the CE dataset is a representative sample of the U.S. population, an increase in the participant sample size for historically marginalized racial groups (e.g., Black and Latinx subgroups) could help with clustering visualizations and early evaluations, thereby allowing for the identification of which populations might benefit from the policy initiative.
Tendentious Data
Do the data represent the subjectivity or impartiality of humans? How does this bias affect the intended outcomes?
Considering that there were qualitative factors inherent to the data collection (e.g., customer interviews and diary entries), the coding and interpretation of these data points involve human subjectivity. The data may therefore embed biases regardless of the protocols used for CE integration. Given this, our interpretation of the data could differ from the original context.

Algorithmic Bias
Harms of Identity Proxy
Could the model treat a particular demographic differently, even without explicit identity markers?
Although the clustering examples were generated twice—without and with race—it must be considered that other variables can serve as a meaningful substitute for race in this context. For example, income and geographic location might generate results that mirror racial-driven outcomes, consequently serving as a proxy for race and vice versa.
Harms of Subpopulation Difference
Are algorithmic outcomes disparate across respective subgroups?
Robust subgroup analysis was not conducted in our clustering application. It may be prudent to evaluate how the clusters might change if only three racial groups were presented (e.g., Black, Latinx, and white) rather than the six racial groups used.
Harms of Misfit Models
If the models are predictive, have you examined their accuracy by subpopulation to ensure performance is not significantly different? Specifically, what is your value orientation and what are the public/social implications of this work?
As was shown in the visualization differences of Figures 10.3 and 10.4, MATLAB-PCA clustering displayed a difference when race was and was not applied, whereas MATLAB-PCA scattering did not. This is important given the same foundational dataset was used. It remains unclear why this was the case. This indicates a need for further investigation to avoid unintended public/social effects.

Human Intent
Do No Harm
What are your goals and intended outcomes? Is any ill intent involved?
The policy goal is to equitably provide subsidies to more sustainable and healthy food options for marginalized populations with negligible impacts to their less marginalized counterparts (e.g., high socioeconomic status or white subgroup). There is no ill intent involved.
Harms of Ignorance
What are the unintended consequences of your work? How can your results be manipulated to abuse or harm?
If the clustering data are further investigated, they could have the unintended effect of informing a counterproductive policy. That is, the identification of these marginalized subgroups could be used to systematically intensify these subgroups' challenges with accessing more sustainable and healthy foods by increasing the costs of such food in their respective communities to offset the cost of healthy foods in less marginalized communities. Again, this is the opposite of our intent but worth highlighting for clarity in future application and interpretation.

*These are only a sample of the reflections that may be relevant when applying the Wells-DuBois protocol. Further considerations are required when attempting to employ systemic equity.

TABLE 10.2 Socioecological Application of the Wells Du-Bois Protocol*

As shown earlier, ML applications such as clustering can be effective in identifying specific relationships among several variables. However, understanding the foundational dataset and model implications is vital when attempting to achieve systemic equity. Otherwise, such applications may perpetuate bias and inequities rather than mitigating them.

10.4.1   Section Overview

In this section, the Wells-DuBois protocol was used on an ML-clustering application involving U.S. food expenditures and environmental impact. This was done to evaluate its bias or inequity given an associated public policy proposal. The checklist for the Wells-DuBois protocol was applied in detail, providing answers for each of its embedded questions. This section provides guidance on how to evaluate the bias or inequity of AI and ML datasets in socioecological systems using the Wells-DuBois protocol.

10.5   Discussion and Future Directions

10.5.1   Other Clustering Activities to Help Achieve Systemic Equity

The clustering example showed the importance of normalizing equitable practices in socioecological AI/ML applications. There is potential in expanding the ML-clustering applications, among other ML applications, to further identify biases. One clustering case could analyze historical data that explores patterns of bias over several years of economic inflation across sociodemographic subgroups to better understand and forecast how these groups might behave in the future. Another could investigate inequities by systematically tweaking a few key data points (e.g., geographic location and socioeconomic status). Undergoing socioecological analyses such as these require the effective administration of tools such as the Wells-DuBois protocol to ensure matters of systemic equity are properly addressed.

10.5.2   Further Discussion on Socioecological Systems

By leveraging the Wells-DuBois protocol to progress toward systemic equity, the three core forms of equity (i.e., distributive, procedural, and recognitional) are more likely to be either addressed or acknowledged for AI/ML tool applications in socioecological systems. As was highlighted in this chapter's clustering application, the nexus of food, energy, and water is an ideal socioecological system in this regard. Competition exists within and across the production sectors of food and energy for water resources (D'Odorico et al., 2018), and there is a need for advanced techniques that support the equitable planning and partitioning of these limited resources (Bozeman et al., 2020; Bozeman, Chopra et al., 2022). Planning for a globally equitable distribution of resources means that we must consider all actors and consumers of food, energy, and water (Bozeman et al., 2019).

As populations increase, agriculture and energy technologies advance, and environmental degradation continues, the need for natural resources will shift across all spatial scales (e.g., locally, regionally, and globally). It is imperative that these shifts are captured across sociodemographic subgroups when collecting and using data to build intelligent and predictive models (e.g., applying clustering in ML). Otherwise, we may distribute resources inequitably while also overlooking more general needs or demands.

For example, a global nutrition transition is the phenomena that explains the changes in diets due to shifting social conditions, such as rising incomes, leading populations to adopt more variety in their diets (Ghattas, 2014). Considering these consumption changes, we must leverage AI/ML tools in a manner that enables decision makers to be both predictive and fair when allocating these scarce resources. It is key that recognitional equity—an under-researched tenet of systemic equity—is embedded in this process to effectively address the needs of the systemically marginalized (Bozeman, Nobler et al., 2022). One must evaluate whose perspectives influence decisions and outcomes, such as ensuring that high-income and high-consumption nations are not conducting resource planning to their benefit alone.

10.5.3 Other Benefits of Employing the Wells-DuBois Protocol and Systemic Equity

Not only does the Wells-DuBois protocol help to achieve equitable practices, but it can also save on costs and protect the value of investments. The cost of AI/ML administration can vary due to factors such as energy needs for supercomputing operation. Nonetheless, these costs do rise as the models are scaled up and become more sophisticated. This can be exacerbated if expensive and time-consuming rework is needed to adjust for emergent biases or inequities.

Equity-centered frameworks and tools such as the Wells-DuBois protocol and systemic equity should be widely adopted, and not just by those explicitly invested in progressing equitable outcomes. It is prudent to encourage and teach equity-centered behaviors when training future researchers and practitioners.

Furthermore, the integration of equity-centered frameworks and tools must migrate from a supplementary practice to a normalized and standard practice for all who perform relevant data techniques. As shown in this chapter, inequity or bias can manifest when data, AI/ML tools, and their outcomes are left unchecked. It is therefore imperative that a culture of systemic equity and bias mitigation be integrated into routine practices across disciplines.

10.5.4 Section Summary

In this section, there was further discussion on the nuances of the concepts and tools explained previously (e.g., AI/ML, the systemic equity framework, and the Wells-DuBois protocol). Insights into the future concept and research directions were also provided. The learning objective of this section is to remember that there is a need to further refine systemic equity approaches in AI/ML applications.

10.6 Chapter Summary

In this chapter, AI/ML bias and inequity in socioecological systems were overviewed, an equity-centered framework (i.e., the systemic equity framework) and tool (i.e., the Wells-DuBois protocol) were highlighted, and the Wells-DuBois protocol was applied to a clustering application to show how equity-centered practices can help achieve systemic equity. It was argued that all AI/ML models that have socioecological implications require standardized assessment for bias. A CE survey dataset was used to illustrate this need. Specifically, it was shown that clustering administration can vary in revealing apparent bias even when using the same foundational dataset(s) and study aim(s). This was followed by an example application of the Wells-DuBois protocol given an associated public policy proposal. The main takeaway is that equity-centered frameworks and tools must be systematically integrated into AI/ML applications regardless of how fair or biased the model components may seem.

Chapter Glossary

Term	Definition
Socioecological systems	A system that involves both social and ecological components.
Systemic equity	A comprehensive framework that helps to mitigate inequitable outcomes and can only be achieved when distributive, procedural, and recognitional equities are addressed simultaneously and over an extended period.
Wells-Dubois protocol	A checklist and tool made up of seven components and three categories that helps to achieve systemic equity by identifying and assessing biases in AI/ML applications.

References

Akter, S., Dwivedi, Y. K., Sajib, S., Biswas, K., Bandara, R. J., & Michael, K. (2022). Algorithmic bias in machine learning-based marketing models. Journal of Business Research, 144, 201–216. https://doi.org/10.1016/j.jbusres.2022.01.083

Andrejevic, M., & Selwyn, N. (2020). Facial recognition technology in schools: Critical questions and concerns. Learning, Media and Technology, 45(2), 115–128. https://doi.org/10.1080/17439884.2020.1686014

Balayn, A., Lofi, C., & Houben, G. J. (2021). Managing bias and unfairness in data for decision support: A survey of machine learning and data engineering approaches to identify and mitigate bias and unfairness within data management and analytics systems. The VLDB Journal, 30(5), 739–768. https://doi.org/10.1007/s00778-021-00671-8

Ben-Hur, A., & Guyon, I. (2003). Detecting stable clusters using principal component analysis. Methods in Molecular Biology, 224, 159–182. https://doi.org/10.1385/1-59259-364-x:159

Borana, J. (2016). Applications of Artificial Intelligence & Associated Technologies. Proceeding of International Conference on Emerging Technologies in Engineering, Biomedical, Management and Science.

Bozeman, J. F., Ashton, W. S., & Theis, T. L. (2019). Distinguishing environmental impacts of household food-spending patterns among U.S. demographic groups. Environmental Engineering Science, 36(7), 763–777. https://doi.org/10.1089/ees.2018.0433

Bozeman, J. F., Bozeman, R., & Theis, T. L. (2020). Overcoming climate change adaptation barriers: A study on food–energy–water impacts of the average American diet by demographic group. Journal of Industrial Ecology, 24(2), 383–399. https://doi.org/https://doi.org/10.1111/jiec.12859

Bozeman III, J. F., Chopra, S. S., James, P., Muhammad, S., Cai, H., Tong, K., Carrasquillo, M., et al. (2022). Three research priorities for just and sustainable urban systems: Now is the time to refocus. Journal of Industrial Ecology, 27(2), 382–394. https://doi.org/https://doi.org/10.1111/jiec.13360

Bozeman, J. F., Nobler, E., & Nock, D. (2022). A path toward systemic equity in life cycle assessment and decision-making: Standardizing sociodemographic data practices. Environmental Engineering Science, 39(9), 759–769. https://doi.org/10.1089/ees.2021.0375

Brey, P. (2004). Ethical aspects of facial recognition systems in public places. Journal of Information, Communication and Ethics in Society, 2(2), 97–109. https://doi.org/10.1108/14779960480000246

Buolamwini, J., & Gebru, T. (2018). Gender shades: Intersectional accuracy disparities in commercial gender classification. Proceedings of the 1st Conference on Fairness, Accountability and Transparency, Proceedings of Machine Learning Research. (pp. 77–91).

Das, S., Dey, A., Pal, A., & Roy, N. (2015). Applications of artificial intelligence in machine learning: Review and prospect. International Journal of Computer Applications, 115(9), 31–41.

Daugherty, P. R., Wilson, H. J., & Chowdhury, R. (2019). Using artificial intelligence to promote diversity. MIT Sloan Management Review, 60(2), 1.

de Ayala, A., Foudi, S., Solà, M. del M., López-Bernabé, E., & Galarraga, I. (2020). Consumers' preferences regarding energy efficiency: A qualitative analysis based on the household and services sectors in Spain. Energy Efficiency, 14(1), 3.

D'Odorico, P., Davis, K. F., Rosa, L., Carr, J. A., Chiarelli, D., Dell'Angelo, J., Gephart, J., et al. (2018). The global food-energy-water nexus. Reviews of Geophysics, 56(3), 456–531.

Donati, F., Dente, S. M. R., Li, C., Vilaysouk, X., Froemelt, A., Nishant, R., Liu, G., et al. (2022). The future of artificial intelligence in the context of industrial ecology. Journal of Industrial Ecology, 26(4), 1175–1181.

Everitt, B., Landau, S., Leese, M., & Stahl, D. (2011). Cluster analysis. Wiley.

Florea, A. R., & Roman, M. (2021). Artificial neural networks applied for predicting and explaining the education level of Twitter users. Social Network Analysis and Mining, 11(1), 112.

Galaz, V., Centeno, M. A., Callahan, P. W., Causevic, A., Patterson, T., Brass, I., Baum, S., et al. (2021). Artificial intelligence, systemic risks, and sustainability. Technology in Society, 67, 101741.

Ghattas, H. (2014). Food Security and Nutrition in the context of the Global Nutrition Transition. Rome, Italy: Food and Agriculture Organization.

"Google apologises for photos app's racist blunder." (2015, July 1). BBC News, 2015/07/01, 2015, Technology.

Green, B. P. (2018). Ethical reflections on artificial intelligence. Scientia et Fides, 6(2), 9–31.

Howard, G. R. (2016). We Can't Teach What We Don't Know: White Teachers, Multiracial Schools. Teachers College Press.

Huang, W., Zhang, Y., & Zeng, W. (2022). Development and application of digital twin technology for integrated regional energy systems in smart cities. Sustainable Computing: Informatics and Systems, 36, 100781.

Jaadi, Z. (2021). A Step-by-Step Explanation of Principal Component Analysis (PCA). builtin.com.

Krause, H. (2017). The Data Equity Framework. We All Count. https://weallcount.com/the-data-process/

Lam, B. S. Y., & Choy, S. K. (2019). A trimmed clustering-based l1-principal component analysis model for image classification and clustering problems with outliers. Applied Sciences, 9(8), 1562.

Leavy, S. (2018). Gender bias in artificial intelligence: the need for diversity and gender theory in machine learning. Journal of Artificial Intelligence Research, 7(3), 112–125.

Li, J., Luo, W., Wang, Z., & Fan, S. (2019). Early detection of decay on apples using hyperspectral reflectance imaging combining both principal component analysis and improved watershed segmentation method. Postharvest Biology and Technology, 149, 235–246.

Miller, J. M., Cullingham, C. I., & Peery, R. M. (2020). The influence of a priori grouping on inference of genetic clusters: Simulation study and literature review of the DAPC method. Heredity, 125(5), 269–280.

Monroe-White, T. (2021). Emancipatory data science: A liberatory framework for mitigating data harms and fostering social transformation. Proceedings of the 2021 on Computers and People Research Conference, Virtual Event, Germany.

Monroe-White, T. (2022). Emancipating data science for Black and Indigenous students via liberatory datasets and curricula. IASSIST Quarterly, 46(4).

Monroe-White, T., & Lecy, J. (2022). The Wells-Du Bois protocol for machine learning bias: Building critical quantitative foundations for third sector scholarship. VOLUNTAS: International Journal of Voluntary and Nonprofit Organizations. 34(1), 170–184.

Moshfegh, A. J., Rhodes, D. G., Baer, D. J., Murayi, T., Clemens, J. C., Rumpler, W. V., Paul, D. R., et al. (2008). The US Department of Agriculture Automated Multiple-Pass Method reduces bias in the collection of energy intakes. The American Journal of Clinical Nutrition, 88(2), 324–332.

Murray, S. G., Wachter, R. M., & Cucina, R. J. (2020). Discrimination by artificial intelligence in a commercial electronic health record—A case study. Health Affairs Blog. https://www.healthaffairs.org/content/forefront/discrimination-artificial-intelligence-commercial-electronic-health-record-case-study

Nadimpalli, M. (2017). Artificial intelligence risks and benefits. International Journal of Innovative Research in Science, Engineering and Technology, 6(6).

Onile, A. E., Machlev, R., Petlenkov, E., Levron, Y., & Belikov, J. (2021). Uses of the digital twins concept for energy services, intelligent recommendation systems, and demand side management: A review. Energy Reports, 7, 997–1015.

Owsiński, J. W., Ciurea, C., & Stańczak, J. (2022). Analyzing cultural biases through the 'reverse clustering' approach: The reality and the interpretation. Procedia Computer Science, 199, 1309–1317.

Özyüksel Çiftçioğlu, A., & Naser, M. Z. (2022). Hiding in plain sight: What can interpretable unsupervised machine learning and clustering analysis tell us about the fire behavior of reinforced concrete columns? Structures, 40, 920–935.

Pannucci, T. E., Thompson, F. E., Bailey, R. L., Dodd, K. W., Potischman, N., Kirkpatrick, S. I., Alexander, G. L., et al. (2018). Comparing reported dietary supplement intakes between two 24-hour recall methods: The automated self-administered 24-hour dietary assessment tool and the interview-administered automated multiple pass method. Journal of the Academy of Nutrition and Dietetics, 118(6), 1080–1086.

PEAR. (2022). Implementing the Washington State Pro-Equity Anti-Racism (PEAR) Plan & Playbook. Edited by State of Washington Office of Governor Jay Inslee. Retrieved from https://www.hca.wa.gov/about-hca/who-we-are/pro-equity-anti-racism-pear

Rodriguez, M. Z., Comin, C. H., Casanova, D., Bruno, O. M., Amancio, D. R., Costa, L. D. F., & Rodrigues, F. A. (2019). Clustering algorithms: A comparative approach. PLoS ONE, 14(1), e0210236.

Steinfeldt, L., Anand, J., & Murayi, T. (2013). Food reporting patterns in the USDA automated multiple-pass method. Procedia Food Science, 2, 145–156.

Tae, K. H., Roh, Y., Oh, Y. H., Kim, H., & Whang, S. E. (2019). Data cleaning for accurate, fair, and robust models: A big data - AI integration approach. In Proceedings of the 3rd International Workshop on Data Management For End-To-End Machine Learning (pp. 1–4).

Turner Lee, N. (2018). Detecting racial bias in algorithms and machine learning. Journal of Information, Communication and Ethics in Society, 16(3), 252–260.

Tyagi, N. (2021). 5 clustering methods and applications. Retrieved from https://www.analyticssteps.com/blogs/5-clustering-methods-and-applications

USBLS. (2022). Annual consumer expenditure surveys 2021. In Consumer Expenditure Surveys. Edited by U.S. Department of Labor.

Weerasuriya, A. U., Zhang, X., Wang, J., Lu, B., Tse, K. T., & Liu, C. H. (2021). Performance evaluation of population-based metaheuristic algorithms and decision-making for multi-objective optimization of building design. Building and Environment, 198, 107855.

End of Chapter Problems

1. Which two system components are represented in socioecological systems?

   a. Symbiotic and ecological components

   b. Symbiotic and environmental components

   c. Social and ecological components

   d. Social and emergent components

2. List the three core equities required to meet *systemic equity*.

3. Define each of the three core equities required to meet *systemic equity*.

4. Which of these options best describes the Wells-DuBois protocol?

   a. A checklist and tool made up of three components and seven categories that helps to identify and evaluate biases in AI/ML applications

   b. A checklist and tool made up of seven components and three categories that helps to achieve systemic equity by identifying and assessing biases in AI/ML applications

   c. A protocol that specifically evaluates the equitability of previous AI/ML applications

   d. A checklist and tool made up of seven components and three categories that helps to identify and evaluate AI/ML computer processing efficiencies

5. List the three categories of the Wells-DuBois protocol.

6. Which three of the seven components fall under the algorithmic bias category?

   a. Harms of misfit models, harms of identity proxy, and do no harm

   b. Do no harm, harms of subpopulation difference, and harms of misfit models

   c. Harms of ignorance, do no harm, and tendentious data

   d. Harms of misfit models, harms of identity proxy, and harms of subpopulation difference

7. Which two of the seven components fall under the human intent category?

    a. Do no harm and harms of ignorance

    b. Harms of ignorance and harms of misfit models

    c. Inadequate data and do no harm

    d. Harms of identity proxy and harms of subpopulation difference

8. Which two of the seven components fall under the bad data category?

    a. Inadequate data and exploitative data

    b. Tendentious data and inadequate data

    c. Inequitable data and implicit data

    d. Inadequate data and biased data

9. Given the example application of the Wells-DuBois protocol, which of these options best describes the response to the tendentious data component (*Note: Refer to Table 10.2*)?

    a. There was definitively no bias or subjectivity within any associated data activity.

    b. The interpretation of the associated data might differ from the original context due to qualitative factors inherent to data collection.

    c. The interpretation of the associated data might differ from the original context due to the biased collection practices of the U.S. federal government.

    d. It is impossible for any system to collect data on socioecological activities while qualifying for bias and inequities.

10. Your job has provided you the results of a market study. They want you to come up with a strategic plan to increase the sales of their nutritious and ecofriendly foods. The study states that those from lower-income households much prefer to purchase sugary, less healthy foods compared to their middle- and high-income counterparts. This same study finds that those from high-income households prefer to purchase more nutritious and ecofriendly foods, comparatively. You also know that this study was modeled from data collected by a group of researchers who largely come from high-income households and that their dataset undersampled lower-income participants. Given this information, respond to the harms of ignorance component of the Wells-DuBois protocol in one to three sentences.

11. Given the socioecological examples provided in this chapter, are there other socioecological systems that could be affected by AI/ML applications? Please provide at least two socioecological examples that affect your everyday life that differ from what was provided in this chapter or expound on one example that was referred to in this chapter by describing how inequitable or bias outcomes might affect your life personally. (Note: There may be good examples from other sections of the book that could provide inspiration.)

CHAPTER 11

Community Engagement for Machine Learning

Bavisha Kalyan
University of California, Berkeley

Anthony Diaz
Newark Water Coalition

Maya Carrasquillo
University of California, Berkeley

Question: Where and how can we incorporate different perspectives in the design of machine learning?

Learning Objectives

Upon completion of this chapter, the student should be able to

- Concisely define the following key terms: environmental justice, community, community-based participatory research, stakeholder
- Design machine learning (ML) systems in partnership with local communities to address context-specific environmental and health engineering challenges
- Identify and implement various tools and methods for community engagement throughout the entire data science tool design process
- Critically assess and evaluate data science projects to create channels for inclusive input from others throughout the design process
- Before any data-gathering activities begin, identify, implement, and review ethical data collection practices that respect the rights and privacy of participants

Chapter Overview

The chapter will explain how community engagement can be used throughout designing ML, emphasizing the importance of involving diverse stakeholders to avoid bias and ensure equitable decision making. Section 11.2 highlights the principles driving community engagement, such as environmental justice (EJ), stressing the need to consider vulnerable populations in ML processes. Methods will be explained for collaborating closely with stakeholders, including community members, to address concerns and achieve social impact. It is linked to community-based participatory research approaches, where citizens are actively involved in data collection, analysis, and advocacy. Stakeholders, particularly community members, must be included in the decision-making process to understand the dimensions of the problem and achieve meaningful results. Section 11.3 will go through the ML design process and provide methods for inclusive engagement at every stage. Section 11.4 presents a case study of community engagement where a community-based organization was supported by researchers to collect primary data on lead exposure in Newark, New Jersey. Overall, the chapter aims to demonstrate how including communities who are affected the most, particularly vulnerable populations, in the ML decision-making process is imperative to more equitable, inclusive, and nuanced conversation about what can go wrong with bias in ML—and what must go right!

11.1   Introduction: Principles and Components of Community Engagement

11.1.1   Prerequisite Knowledge and Context: Case Study of Flint, Michigan

What comes to mind when you think of the Flint Water Crisis? The Flint Water Crisis made national headlines in 2015 when the tap water in Flint, Michigan, homes came out of the tap brown. People were alarmed at the color of the water and were upset to learn that children in Flint have record levels of toxic lead metal in their blood. And so the question was asked: Was the brown water related to the blood lead levels in the children?

Community organizers, scientists, citizen scientists, and researchers all contributed to collecting data and bringing political and media attention to the water in Flint. After investigation, they discovered that water was not only brown, it contained lead. The Flint water source was switched from the Detroit Water and Sewerage Department to the Flint River as a cost-saving measure. However, the Flint River water was not properly treated, leading to corrosion of the city's aging lead pipes. As a result, lead began leaching into the water supply, contaminating the drinking water. Lead pipes were widely used in many water systems until 1986, including in Flint.[1] Lead is invisible, but the brown water in taps serves as a warning signal of corrosive water that is dissolving different metals and minerals in the pipes, such as copper, and coloring the water brown (Clark, 2018).

---

[1]Lead pipes were banned by the 1986 Safe Drinking Water Act Lead Ban.

The lead poisoning in Flint was particularly devastating because the population of Flint were a majority of low-income and marginalized racial groups; Flint has been a historically predominantly Black city. According to the U.S. Census data from 2010, the racial composition of Flint was approximately 53 percent Black, 37 percent white, and 4 percent Hispanic or Latino. For several decades, Flint also struggled with high poverty rates significantly above the national average. The poverty rate is the percentage of the population living below the federal poverty line of $12,880 in annual income. In 2010, the poverty rate in Flint was approximately 40 percent, meaning that about 40 percent of the city's residents lived in poverty.

The racial demographics and poverty rates of the population in Flint made the city particularly vulnerable during the Flint Water Crisis. The high poverty rate meant that many residents faced economic challenges, limiting their ability to access and pay for alternative sources of water and other basic resources and services. The crisis and its handling raised concerns about environmental injustice. The city's water source was switched without accounting for the potential health risks and consequently affected the city's low-income and Black residents. These residents, in particular, already face high health risks due to their socioeconomic conditions and environmental exposures, and lead poisoning and related health issues can have long-term effects, compounding the disadvantages. Lastly, years of disinvestment in Flint's infrastructure created a city at risk for harm.

The crisis created deep mistrust between the community and government, as many residents and activists felt that their concerns were dismissed, resulting in a breakdown in communication. Activists, researchers, and community leaders worked to raise awareness of the crisis and pushed for more comprehensive solutions. Once lead from the pipes starts leaching into the water, water systems have an option to replace the lead pipes—if they can be found. Since the pipes were installed so long ago, many cities have old, paper, incomplete records and need to physically verify if there are lead pipes. Finding lead pipes requires excavating the ground outside of a home to verify if the pipe is made of lead. If it is made of lead, the pipe can be pulled out of the ground. Alternatively to find the pipes, we could test the water to see if lead is leaching into the water from the pipes. If the water contains lead, then there might be a water pipe containing lead.

Predicting the likelihood of a house containing lead pipes is an opportunity for ML. Given a correlation between the age of the homes, the presence of lead pipes, lead poisoning in children (captured by mandated blood lead level testing), and historical disinvestment of infrastructure in some parts of the city, there have been attempts to build ML models to detect lead hotspots. However, lead poisoning is not an isolated occurrence. Many actors have played a role in leading to historical harm of people who continue to face harm. To prevent bias and a limited perspective of data scientist modelers, we must engage with multiple stakeholders when designing ML models.

In the Flint Water Crisis, and throughout the subsequent actions taken to remove the lead pipes, including developing ML algorithms to predict locations of lead pipes, the importance of community engagement became evident. Community engagement can give a voice to the affected communities. Many community members in Flint had raised warnings about the water quality but were initially dismissed. Engaging with the community empowers them to share their experiences, concerns, and knowledge, ensuring that their perspectives are considered in decision-making processes. Community members knew firsthand the health issues they were facing due to the contaminated water and can help understand the full extent of the problem and tailor solutions to address those specific needs. The crisis eroded community trust in government officials and institutions responsible for public health, and engaging with the community in a transparent and collaborative way can help rebuild trust and foster a more positive political environment. Community engagement also offers an opportunity to involve different stakeholders, including government agencies, nonprofit organizations, and researchers; resources can be pooled together to address the immediate needs and to prevent similar crises from happening in the future.

The Flint Water Crisis was a complex issue with multiple contributing factors. Lead pipes and the subsequent toxic contamination of water is an example where human health, ecosystem health, and infrastructure all overlap. The response and recovery efforts have involved various stakeholders working together to address the challenges faced by the community and provide valuable lessons to prevent similar incidents in the future. These lessons can be adapted to how we approach data science problems where community engagement is necessary. This chapter will introduce how to build ML models with multiple stakeholders—particularly in topics of health and the environment.

Community engagement emphasizes close collaboration among stakeholders when tackling a concern. Community engagement provides an approach to collaborative problem solving, which includes projects where citizens or community members collect, analyze, and engage with data. This section discusses how community engagement is integral in the steps toward social impact.

To contextualize the importance of community engagement, let's now explore how these principles are deeply intertwined with the concept of EJ. EJ, a response to the unequal environmental burdens faced by marginalized communities, shares a strong connection with community engagement. By understanding the historical context and objectives of EJ, we can better grasp the importance of community-driven approaches and their potential for promoting social impact and empowerment. Through a community engaged process, the community will be further invested in the project if they have a deeper understanding of the solution and the work likely to sustain after the data scientist has moved on to another project.

11.1.2   Brief Introduction to Environmental Justice and Environmental Data Justice

Research shows that people who face social and economic disadvantages are more likely to experience health issues and even higher rates of illness and death due to both harmful chemicals in the environment and various social factors affecting health, including things like poverty, racism, lack of job opportunities, and more. These social factors worsen people's well-being, and when combined with environmental health threats, the risks to these populations become even greater.

For example, certain communities may be exposed to elevated levels of harmful substances like lead, and at the same time, they may also struggle with limited access to essential resources like food, health care, and safe housing. This combination of environmental hazards and social challenges puts these communities at a higher vulnerability for health problems.

This unequal distribution of health burdens and the underlying social structures that perpetuate persistent environmental issues and social segregation is at the heart of EJ.

EJ formally began as a movement, emerging from the organized civil rights movement in the preceding decades. The EJ movement emerged as a response to the disproportionate exposure to pollutants and environmental hazards faced by marginalized and racialized communities. The EJ movement emphasized that a healthy environment, a place "where we live, work, and play," was a necessary component of a healthy life. The movement sought to ensure the access, maintenance, and assurance of a healthy environment for all people regardless of their background (Holifield, 2001).

EJ is strongly linked with Environmental Hazard Evaluation (EHE). Two instances of landmark EJ moments occurred in 1982 when a majority Black community mobilized against a polychlorinated biphenyl (PCB)–contaminated landfill in Warren County, North Carolina, and in 1987 when the United Church of Christ Commission for Racial Justice published the report, "Toxic Wastes and Race in the United States," documenting the correlation between race and location of toxic waste facilities. The 1991 First National People of Color Environmental Leadership Summit broadened the principles of EJ from focusing on disproportionate exposures to environmental toxins and pollution to include other social or quality-of-life issues, such as housing, transportation, employment, food access, inclusion, empowerment, and others. They recognized environmental injustice was created by "over 500 years of colonization and oppression' using violence, genocide, and bodily harm" (Vera et al., 2019). Efforts such as these brought change to U.S. federal laws and policies, including a 1994 Executive Order by President Bill Clinton and subsequent establishment of the Environmental Protection Agency's (EPA's) Office of Environmental Justice. More recently, the Obama administration issued the 2011 Memorandum of Understanding, and the Biden administration's Justice 40 initiative reinvigorates federal agency responsibility for achieving EJ.

Despite these efforts, environmental injustice and environmental racism are perpetuated through U.S. environmental policies and infrastructure that sustain permission-to-pollute systems.[2] Environmental racism implicates the state as complicit in harm, often through insufficient and damaging regulations and laws, lack of funding, relationships with industries responsible for pollution and hazards, or being the party responsible for pollution. Environmental racism is inseparable from EJ, and understanding environmental racism explains the origins and motivations of the EJ movement. Scholars have emphasized critical and intersectional approaches to EJ frameworks that view environmental racism as part of the matrix of domination. The matrix of domination describes the overall systematic distribution of power in society, while intersectionality is used to understand a specific social location of an identity using mutually and intersecting constructing features of oppression (Vera et al., 2019). Any data science solutions created for problems with or related to EJ must consider how the problem addressed results from multiple historic oppressive structures, and any solutions created affect communities differently based on their specific location and identity. Community engagement is one option for co-designing data science systems to address critical societal issues, alongside empowering local communities to catalyze social impact.

---

[2]Permission-to-pollute systems encompass the practice of industries being allowed to release pollutants into the environment as long as the pollutant is regulated by permits or under a regulatory contaminant level.

Designing a data science solution requires understanding the problem from the perspective of the stakeholders affected. To understand if a problem can be addressed with an ML solution, one needs to talk to stakeholders to understand all the dimensions of the problem. In environmental and social topic areas, community members are usually key stakeholders and are also most affected by the adversarial health and environmental effects. Therefore, we want to include these most affected stakeholders in the decision-making process throughout the ML design. Community engagement is the collective term we will use to describe how to engage community members.

11.1.3 Data Justice

Data justice is distinct from, but overlaps with and is inspired by, environmental and social justice (ESJ). Data justice refers to the fair and equitable use of data, ensuring that data collection, analysis, and interpretation do not perpetuate or exacerbate existing social inequalities, discrimination, or power imbalances (Vera et al., 2019). Data can be used for environmental or social justice causes; however, this is separate from the concept of data justice. Data justice means stakeholders, especially those marginalized by environmental and health impacts, have the possibility to understand, challenge, and be part of algorithmic decisions (Dencik & Sanchez-Monedero, 2022). It emphasizes the need to highlight ethical, social, and political issues associated with data, particularly in the context of data-driven technologies, algorithms, and artificial intelligence systems.

The concept of data justice recognizes that data have become a critical aspect of modern life, influencing decision making in various domains, including health care, education, employment, law enforcement, and public policy. Data-driven tools have become part of an integrated social and environmental justice agenda, but these problems and solutions may not actually be about data (Dencik & Sanchez-Monedero, 2022; Vera et al., 2019). Nonetheless, data-driven tools themselves can also lead to unintended negative consequences, such as privacy violations, biased algorithms, and unfair profiling. Data justice asks us to not overlook the underlying conditions that produce an uneven distribution of power and resources between different groups in society and consider if the data tool reinforces power imbalances (Constanza-Chock, 2020; Vera et al., 2019).

Data justice calls for:

Transparency: Making data processes, methodologies, and algorithms transparent and understandable to the affected communities to foster accountability and trust.

Privacy and Consent: Respecting individual privacy rights and obtaining informed consent from data subjects when collecting and using their data.

Avoiding Bias: Identifying and mitigating biases in data collection and analysis to prevent unjust discrimination and harmful effects on vulnerable populations.

Empowerment: Enabling individuals and communities to have control over their data and the ability to influence how it is used for decision making.

Accountability: Holding organizations and institutions responsible for the impact of their data practices and ensuring they address any negative consequences.

Data justice is essential to promote social and environmental justice and protect human rights. Data justice seeks to create a more equitable and inclusive data ecosystem where data-driven technologies do not harm vulnerable or marginalized groups. The design process, including the relationship between the data scientist and stakeholder, needs to incorporate principles of justice, which can start with building a design process centered on community engagement (Costanza-Chock, 2018).

11.1.3.1   *Examples of Data Justice*

To advance and draw attention to data justice, activist data projects are incorporating traditionally unheard perspectives into the data design process. A group called Data 4 Black Lives is moving beyond data projects that document harm toward creating data projects that use data science to create concrete and measurable change in the lives of Black people. (http://d4bl.org; Drake, 2016). They argue that data systems can empower communities of color but have historically been used as a tool for discrimination and oppression.

Resonating with this critical perspective, the Detroit Digital Justice Coalition formed in 2009 to address the digital divide within Detroit and generate knowledge with community members around use, rights, and ownership of technology. Their principles of "digital justice," inspired by EJ principles, include access, participation, common ownership, and healthy communities (https://www.alliedmedia.org/ddjc/principles).

Another example is the Indigenous Data Sovereignty movement, which advocates for the right of a nation to govern the collection, ownership, and application of its own data (Kukutai & Taylor, 2016). The United States Indigenous Data Sovereignty Network helps ensure that data for and about Indigenous nations and peoples in the United States (American Indians, Alaska Natives, and Native Hawaiians) are utilized to advance Indigenous aspirations for collective and individual well-being (Rainie et al., 2017).

These projects originated from recognizing and witnessing how data systems can reinforce racism and are part of the emerging call for "environmental data justice," preliminary defined as embracing "public accessibility and continuity of environmental data and research, supported by networked open-source data infrastructure that can be modified, adapted, and supported by local communities." The broad range and aim of these movements uncover new thinking, emergent practices, and techniques that actively combat the extractive nature of data systems and instead offer robust, community-developed data systems.

11.1.4   What Is a Community?

The EJ movement emerged out of an informal network of community organizations, regional, and national groups, centered on empowering the affected communities so that they can work toward solving their specific environmental and health problems (Williams, 1999). The movement reaffirms that communities are both places of cultural identity and sources of meaning within the larger world. Communities are formed from many social dynamics, for example, the pattern of racial segregation in residential housing in the United States created communities with shared experiences of discrimination due to race and ethnicity (Williams, 1999). With the interconnectedness of the Internet, communities can now be connected by virtual spaces.

An appropriate geographic unit should be used to analyze concerns a community faces, such as a political region, a neighborhood, and geographical measures (such as

ZIP codes and census tracts) (Williams, 1999). Each unit has its own advantages and limitations, and a combination of different units can be used to capture the effects faced by the community. Ideally, the data scientists work with the community to determine which unit they consider to be important and then secure adequate data for the area under study (Williams, 1999).

Community members play a vital role in understanding and addressing the complex challenges posed by social and environmental issues. Community members are not only the most affected by health and environmental impacts, but they also possess contextual expertise that is essential for developing meaningful and inclusive data science solutions. By building on the principles of community engagement and its connections to frameworks like participatory research, we can ensure that data science efforts are both ethically grounded and capable of delivering actionable insights with long-lasting social impact (Williams, 1999).

Community engagement is the process of yielding agency to communities throughout a project process so that they can create, use, and disseminate technology, data, and research (Key et al., 2019). In addition, community engagement is a tool through which projects can advance social and policy changes.

11.1.5   Community-Based Participatory Research

Community engagement falls under a broader approach to research, called participatory action research, a framework that emphasizes shared inquiry and action across stakeholders with the goal to democratize the decision-making processes of collective knowledge production and cycles through reflection and action. In participatory action research projects, community engagement begins at the onset of a project and can take on many forms such as consulting with residents throughout the process to working on a community-driven project (Key et al., 2019).

Community-based participatory research (CBPR) is a form of action research where community partners are collaborators on a topic of study, rather than researchers observing from the outside. CBPR actively involves community members in every stage of the research process, from defining the problem to implementing findings (Flicker et al., 2007). Each partner is recognized for their unique strengths. CBPR begins with finding a research topic of importance to the community. The ideal aim of CBPR is to combine knowledge with action to achieve social change and improve health outcomes.

With a CBPR approach, a deeper understanding of a community's unique circumstances emerges and should be central throughout the design and decision-making process. When done properly, CBPR bridges the gap between scientists and communities through the creation of shared knowledge and valuable experiences. A collaboration built through CBPR can lend itself to mutual ownership of culturally appropriate, effective, and efficient processes and products. CBPR involves communities throughout every step of the decision-making and research process, summarized in Table 11.1.

There is a rich body of literature on community engagement approaches for addressing public health and environmental problems. We encourage you to engage with and draw on community engagement approaches as you establish your approach.

11.1.6   Citizen Science and Community Science

When citizens or a community collects data about a project, we broadly call this citizen science or community science. Citizen science projects have been powerful forms of

Phase of Research	Community Engagement in Decision-Making Process
Input	Communities initiate research ideas, generate hypotheses and projects, determine recruitment strategies and distance funding, and form advisory and partnership steering committees
Process	Communities remain engaged throughout data collection, such as conducting surveys and interviews, analysis, and interpretation of results
Outcome	Communities provide avenues for disseminating information, mobile knowledge gained in the project for social change

The **key principles of CBPR** are laid out in Billies et al., (2010), Costanza-Chock (2020), Wallerstein & Duran (2006), and Wallerstein & Duran (2008):

1. Recognizes community as a unit of identity in which all partners have membership

2. Builds on strengths and resources within the community to address local concerns and solve relevant problems

3. Emphasizes democratic partnerships between all project members as collaborators through every stage of knowledge and intervention development

4. Requires a deep investment in change that carries with it an element of challenging the status quo, improving the lives of members in a community, and attending to social inequalities

5. Integrates knowledge and action for mutual benefit of all partners

6. Promotes a co-learning and empowering process that attends to social inequalities and addresses health from both positive and ecological perspectives

7. Involves a cyclical and iterative process in which a problem is identified, solutions are developed within the context of the community's existing resources, interventions are implemented, outcomes are evaluated, and interventions are modified in accord with new information as necessary

8. Promotes project partners' humility and flexibility to accommodate changes as necessary across any part of a project and fosters co-learning and capacity building

9. Disseminates findings and knowledge gained to all partners

10. Involves a long-term commitment by all partners

TABLE 11.1 Community Engagement at Every Level of Decision-Making and Research Process

data that have influenced policy. Community and citizen science projects have yielded informative, neighborhood-level datasets, as well as increased community capacity to collect and assess data. Examples of citizen science projects include the following.

In 1991, Erin Brockovich exposed the existence of a dangerous contaminant polluting a town's groundwater, a toxic hazard that otherwise might have stayed invisible. In 1996, Erin and a law firm won a $333 million settlement against Pacific Gas & Electric for 650 plaintiffs, at the time the largest toxic tort settlement in American history.

While the terms are often interchanged, community science and citizen science can be distinguished. Citizen science is seeing a "rebranding" to community science, as many people contest the exclusionary term citizen science as it can be perceived to exclude those without citizenship status within a given nation (Cooper et al., 2021; Lowry & Stepenuck, 2021). However, citizen science is more commonly used and has been codified into institutions globally. Community-engaged strategies can include both citizen science and community science, but using community science or citizen science datasets for data science projects is not a form of community engagement. A scenario where community science or citizen science is done in collaboration with data scientists is still very rare—see Section 11.4 for an in-depth example of an ongoing project. A scenario where community science or citizen science is *not* a form of community engagement is when a community science or citizen science project is conducted, such as citizens taking photos of insects to create a database, but the decision-making power over the project lies with the academic institution, government agency, or nongovernmental organizations. The dataset is then used by data scientists to build an ML model. While the dataset originated from community science or citizen science, the data science design process was separate.

11.1.7   Section Summary

Community-engaged strategies can shape the project approach to elevate EJ concerns and create long-term improvements in the environment and health of communities. The development of predictive ML models traditionally follows a data scientist–centered approach, in which the data scientist often has more power (in terms of scientific authority and available resources) over local communities, especially underserved communities (Holzmeyer, 2021). This unequal power relationship can cause mistrust between the scientists and the communities and potentially harm the community. For example, data science requires large, detailed datasets that are not necessarily publicly available, while social and environmental community data concerns typically involve multistakeholder conversations in a large and regional sociotechnical system and require transparency and trust.

When done ethically and collaboratively, building datasets and data science tools for decision making through a community-engaged process fosters interagency collaboration, prioritizes vulnerable communities, and increases coordination between research, government, and advocacy actors (Dietrich et al., 2022; Morello-Frosch et al., 2022; Pace et al., 2022). Ideally, together we can design an ML algorithm that intertwines the values of trust across its basic performance and aligned purpose (Vashney, 2022).

11.2   Project Initiation: Steps of Conducting Community-Driven Environmental Data Science

When datasets increase in quantity and complexity, traditional statistical analyses face limitations to understand and describe patterns in the data. ML can overcome traditional statistical limitations because it does not require much prior knowledge about a problem to be able to fit a model to the data, making ML well suited for solving complex data patterns (Zhong et al., 2021). Because ML is characteristically more effective than traditional statistical tools in handling a wide variety of data formats, including text, images, and graphs, where the important information is contained across more than one variable and not known ahead of time, ML is suited for a variety of social and environmental applications. For example, ML can be used for applications such as assessing environmental risks, evaluating the health of water and wastewater infrastructure, optimizing treatment technologies, identifying and characterizing pollution sources, and performing life cycle analysis, among others (Zhong et al., 2021). Overall, the current uses of ML in terms of social and environmental applications are summarized into making predictions, identifying feature importance, anomaly detection, and discovering materials and chemicals, with examples in Table 11.2 (Zhong et al., 2021). The unique properties of ML are especially suitable for solving complex social and environmental problems with rich sets of input features.

Traditional data scientists take a preexisting dataset when creating an ML algorithm to solve a problem. The data scientist should be familiar with the data itself. For example, if a data scientist is developing an ML predictive model to predict contaminated water sources, the data scientist should know the sources of the contamination and how it interacts with the water to know how and if the data are describing the chemical and physical properties of the contamination.

In brief, after the dataset is obtained, the data are preprocessed to clean the data (i.e., standardization and normalization, removing duplicates), and the data are then split between a training set and a testing set. The training set is then used to train an appropriate model. The hyperparameters of the model are tuned and the outcome of the model is checked on the testing set. The metrics (i.e., accuracy, true positive rate) used to evaluate the model are chosen based on the purpose of the model. The model is then taken back to the stakeholders for feedback and then adjusted in an iterative manner.

Now, let's delve into the different stages of community-driven environmental data science and how community engagement can be seamlessly integrated throughout the ML design process to foster mutual trust and collaboration between researchers and communities. As discussed earlier, community engagement emphasizes collaboration and partnership with stakeholders, including community members, to tackle concerns effectively. Community engagement should be introduced throughout the data science

Application of Machine Learning	Example of Application
Making predictions	Predict atmospheric pollutants (such as $PM_{2.5}$) based on past measurements at different locations.
Identifying feature importance	Given that an ML model has satisfactory predictive performance or has already learned the correct underlying relationship, the model can be used for a new understanding. For example, a model built to predict ecological processes, feature extraction, or model interpretability will tell us about the stressors and ecologically important environmental factors that interact and drive ecological processes.
Anomaly detection	Detect credit card fraud by comparing new observations to a historical distribution of observations.
Discovering materials and chemicals	Build a model that provides options for new structures which can be used to develop biopolymers.

Source: (Zhong et al., 2021)

TABLE 11.2 Applications of Machine Learning Used in Social and Environmental Projects with an Example

process following three phases: (1) preparation for modeling, (2) model development, and (3) after modeling. Figure 11.1 describes these phases with the actions taken in each step. Section 11.3 will detail each phase in Figure 11.1.

11.2.1   Phase 1: Building Partnerships

Before designing an ML project, a relationship with the community should be formed. During this phase, building a genuine relationship with the community is important to establish trust, rapport, and respect. In the data scientist–centered approach, it is assumed that the data scientist can make decisions about the data application and interpretation by theoretically placing themselves in the situation of citizens and empathizing with local people's perspectives. However, data scientists who come from different socioeconomic, cultural, gender, religious, regional, institutional, political, racial, and ethnic backgrounds than the community they are working with may have difficulty fully and authentically understanding local people's experiences. Regardless if you have a shared identity or not, this step in the process is critical to not operate with assumptions, but to take the time to build relationships and a critical understanding of the people you are working with. Only by honestly admitting and reflecting on this weakness and recognizing the power inequality between data scientists and local partners can researchers and practitioners truly respect community knowledge and be sincerely open minded in involving local communities—especially vulnerable populations exposed to environmental and health injustices—in the center of the data science design process.

Data scientists need to be with people who are affected by local concerns to co-create ML systems so that the outcomes are valuable and beneficial to the local community. To achieve this, we encourage data scientists to conduct hands-on field science. Creating social impact lies *with* local people and their long-term advocacy. We also encourage data scientists to genuinely collaborate with local people to address pressing social concerns and further immerse themselves in the local context to become "scientific citizens" (Irwin, 2001). A good partnership includes strong community leadership, active participation, diverse skills, effective networking, shared values and power, willingness to challenge entrenched powers, focus on larger contexts, respectful dialogue and critical reflection, and support from broader institutions, finances, and networks (Wilson et al., 2014). One approach is to attend, after obtaining consent and respect, existing community meetings such as task force meetings, steering committees, or other community gatherings.

Consider your existing or future relationships you wish to establish with local partners:

— Why do you want to work with a group or community?

— What assumptions do you have about the community?

— What potential power dynamics exist between you, the institutions or groups you work for or represent, and the community you wish to work with?

— How much time is needed for you to build trust and rapport with a community before collaborating on a data project?

PHASE 1. BUILDING PARTNERSHIPS:

- Establish trust, rapport, and respect
- Honestly admitting and reflecting as the outsider and recognizing the power inequality between data scientists and local partners
- Respect community knowledge and be sincerely open-minded in involving local communities
- Conduct hands-on field science: genuinely collaborate with local people and immerse in the local context

PHASE 2. PREPARATION FOR MODELING - DONE IN CONVERSATION WITH STAKEHOLDERS

Formulating the question
- Understand the system context
- Identify stakeholders and community leaders

Data collection
- Existing datasets
- Experimental or observed data with local communities
- Choosing "representative" data points with stakeholders

Data preprocessing
- Standardization and normalization
- Removing duplicates
- Missing values treatment
- Outlier detection
- Encoding categorical features
- Feature selection and reduction

Data splitting
- Training, validation and test datasets
- Cross-validation on the training and validation datasets
- Data similarity consideration

PHASE 3. MODEL DEVELOPMENT - DETERMINED BY STAKEHOLDER REQUIREMENTS

Evaluation metrics
- Regression: RMSE, MSE, MAE, R^2, etc.
- Classification: Accuracy, F1-score, AUC-ROC

Model training
- ML algorithm selection based on stakeholder desires

Model evaluation
- Performance on the test dataset to meet stakeholder requirements

Model performance enhancement
- Ensemble or stacked model
- Transfer learning
- Domain knowledge modification from stakeholder input

PHASE 4. AFTER MODELING

Model interpretation
- Usability and interpretability of the model by stakeholders

Model Applicability
- Ensure the model fits the context and application

Model Deployment
- Stakeholders retain ownership over the model

FIGURE 11.1 Typical workflow for the development of ML models adapted from Zhong et al. (2021) combined with steps for community engagement in the processes. RMSE, MSE, MAE, and R2 refer to the root mean squared error, mean squared error, mean absolute error, and coefficient determination, respectively, while AUC-ROC means the area under the curve–receiver operating characteristic curve.

11.2.2   Phase 2: Preparation for Modeling: Done in Conversation with Stakeholders

11.2.2.1   *Formulating the Question: Understand the System Context, Identify Stakeholders, and Define the Problem and the Goals*

Defining the problem and goal of data science projects should be done using community engagement. Understanding who the key stakeholders are is a crucial step to developing a strong community engaged project. This section will provide a brief introduction to the available methods to learn from and build with stakeholders, and we encourage readers to delve deeper into these methods.

In addition to lists, a stakeholder map visually allows us to understand who the stakeholders are, what decision-making power they have in a given system, and stakeholder relationships to each other.

When obtaining identifiable information, particularly if you conduct academic research, institutional review board (IRB) approval will have to be obtained. In addition, to collect identifiable information that will be presented back later, privacy and confidentiality are important to protect the participants' information. See Section 2.2.6 for more information on the IRB process.

11.2.2.2   *Identifying Stakeholders, Decision Makers, and Community Leaders*

Stakeholders, including decision makers and community leaders, might span multiple sectors, such as politicians, government agencies, grassroots organizations, activists and advocacy groups, clergy, and industry companies. If you have no context to start from, one option is to start by reading other articles and interviews to determine the names of several influential people and their association with particular organizations and projects. Systematically connect with and interview these stakeholders. In the interview, center the questions on topics such as what data problems are important to the stakeholder, what barriers and challenges they face, and what successful methods they use.

Lists (Table 11.2) or stakeholder maps (Figure 11.2) are effective methods of collecting and displaying stakeholder information. Use the interview to fill out Table 11.3 with the list of the stakeholders as well as their rights, responsibilities, and power that the stakeholder has within the system. Don't forget to ask each interviewee which stakeholders they believe are influential and add them to the list. Repeat this process until no new names or organizations are added to the list.

In addition to interviews, focus groups, workshops, and participant and site observation are ways to gather information as part of the first step in identifying what the community deems a problem. During these information gathering steps,

Fill out the table to evaluate the stakeholder and their rights, responsibilities, and power.

Stakeholder	Rights	Responsibilities	Power
List stakeholders.	What right(s) does the stakeholder have in the system?	What is the stakeholder responsible for in the system?	Describe their influence over the system.

TABLE 11.3   List of Stakeholder Information

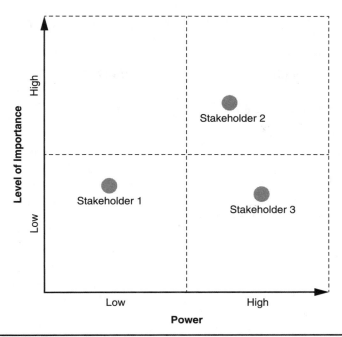

FIGURE 11.2 Stakeholder importance map.

a free-format and open group discussion of problems can generate a long list of concerns and solutions. To develop a clearer picture of the problem, focus on topics to help guide but not dictate the conversation, such as questions about labor and money, capital costs, and what data are already available. Allow community representatives to present their ideas before the data scientist presents an idea. Data scientists can help the community understand the details, limitations, and challenges of a particular data science solution.

Consider how each stakeholder or stakeholder group is unique and has sub-identities:

— If a community-based organization is a stakeholder, consider what population does the community-based organization serve?

— What motivates and incentivizes each stakeholder? What are their barriers? What is their goal?

11.2.2.3 Understanding Stakeholder Relationships
Once the stakeholders are identified, their importance and power can be compared to each other. We can visually represent the relationships on a simple plot with the x-axis ranking the power of the stakeholders (potential impact on the success or outcome of the initiative) and y-axis ranking stakeholders by their level of importance (or interest and involvement) in the decision-making process and outcomes (see Figure 11.2).

Remember that we are comparing stakeholders' power and importance relative to each other, and don't forget to place yourself in the stakeholder map.

Stakeholders can be categorized into different quadrants based on their position on the graph:

Low Power, High Importance These stakeholders have a strong interest in the project's success but have limited influence. They may be important to keep satisfied, but they might not drive major decisions.	**High Power, High Importance** These stakeholders have significant influence and are highly interested in the project's outcome. They are key players and need close engagement and careful management.
Low Power, Low Importance These stakeholders have little influence or interest in the project. They might be monitored to ensure no adverse impact occurs, but their involvement may not be a high priority.	**High Power, Low Importance** These stakeholders have significant influence but may not be as directly interested in the project. They need to be kept informed but may not require as much active involvement.

The stakeholder importance map helps data scientists understand where to focus their efforts in terms of engagement, communication, and decision making to ensure that key stakeholders are appropriately addressed and their needs and concerns are taken into account throughout the ML design process. This approach is particularly useful in complex projects or situations where multiple parties with diverse interests are involved.

To explain and draw relationships between stakeholders, fill out Table 11.3. Table 11.3 is a matrix of the stakeholders (along the rows and columns) and the cells explain the relationships between them. Write down any shared resources between stakeholders and how, or if, they interact with each other.

Determining the stakeholders summary of steps:

→ Read background articles and interviews to determine the names of several influential people and their association with particular organizations and projects.

→ Systematically connect with and interview these stakeholders. Alternative ways to gather stakeholder information include focus groups, workshops, participant observation, and site observation.

→ Fill out Table 11.3 to evaluate the stakeholder and their rights, responsibilities, and power.

→ Create a importance stakeholder map (Figure 11.2) showing the relative importance and power of each stakeholder.

→ Evaluate the relationship between each stakeholder by filling out a stakeholder matrix (Table 11.4).

	Stakeholder 1:	Stakeholder 2:	Stakeholder 3:	...
Stakeholder 1:		Relationship between Stakeholder 1 and Stakeholder 2	Relationship between Stakeholder 1 and Stakeholder 3	...
Stakeholder 2:			Relationship between Stakeholder 2 and Stakeholder 3	...
Stakeholder 3:				...
..				

TABLE **11.4**   Stakeholder Relationships

Consider the following questions when establishing work with community partners:

— Does your community partner have the organizational, training, and fiscal capacity to own and manage the project process? For example, do they have the fiscal and managerial capacity to manage and adhere to federal grant reporting requirements?

— Does your community partner desire the responsibility of acting as principal investigator (PI) or project manager? Are they comfortable with sharing responsibility with their academic partners?

— What is the distribution of the burden of the project activities?

11.2.2.4   Phase 2: Data Collection

Typically, building data science systems with a community requires a tremendous community outreach effort for data collection and analysis. In a hands-on field approach, the data scientist and community partners will be designing the data collection steps that work within the community's capacity and resources but still obtain the validity and spread for rigorous data analytics. Quantitative and qualitative data can be collected. Qualitative methods provide nuances, perspectives, and data about the system and can be used to supplement quantitative methods. Qualitative methods (described in Section 2.2.5) include interviews, surveys, focus groups, ethnographic research, and workshops. One must determine how much data are appropriate to answer the research question and consider the environmental, financial, and labor costs of the data and if the cost of collecting data and the associated risk of collecting a large dataset are worth the outcome of creating the dataset and model (Vera et al., 2019). With your community partners, budget for curation and documentation at the start of a project and only create datasets as large as can be sufficiently documented.

Similarly to most applications of ML, applying ML with community engagement faces challenges. The first challenge stems from data scarcity and data quality. Even when there are data, differences in collection methods due to differences in the environment (such as water quality, soil sediments, and population demographics) can create discrepancies and hinder the creation of a large, consistent, and high-quality dataset for ML. The second challenge arises from overfitting. Overfitting is where a model can have excellent predictive performance on the training samples but fails to accurately predict for new samples. Lastly, traditional statistical tools may be more appropriate than ML in some cases, especially when there are small sample sizes (Flicker et al., 2007). Sometimes, not every problem should be solved by ML tools directly—thoughtful design is needed to address social and environmental problems with ML tools (Holifield, 2001).

11.2.2.5 Brief Introduction to Qualitative Data Collection

Qualitative data are a form of data that can be used to address and understand a problem. Qualitative research is a systematic discovery with the intention of generating knowledge of social events and processes by observing how people react, interact, and interpret the world around them (Montoya & Kent, 2011; Ulin et al., 2004). The qualitative research process is flexible and iterative between design and discovery. Qualitative data can be expressed using participants' words, in images, and sometimes in numbers (Montoya & Kent, 2011; Ulin et al., 2004). Qualitative data can also supplement and/or guide quantitative data collection. This section will briefly describe some of the methods you can use to collect qualitative data. Table 11.5 summarizes some traditional qualitative methods and examples of studies using the methods.

11.2.2.6 Ethical Data Collection and Compliance

When collecting identifiable information, the privacy and confidentiality of the information must be secured and consent has to be obtained. The guidelines provided by an IRB will ensure the data collection process is ethical. Obtaining approval from an IRB is especially important when working with vulnerable populations, such as low-income people, pregnant women, and children.

The IRB process was formed in the aftermath of unethical research practices that took place throughout history, raising the need for ethical oversight and protection of human subjects in research. One example that highlighted the need for ethical oversight in research is the Tuskegee syphilis study.

The intent of the study was to record the natural history of syphilis in Black people. The study was called the "Tuskegee Study of Untreated Syphilis in the Negro Male." When the study was initiated in 1932 there were no proven treatments for the disease. A total of 600 men were enrolled in the study. From this group, 399 men who had syphilis were a part of the experimental group and 201 men were assigned as control subjects. Most of the men were poor and illiterate sharecroppers from the county. For their participation in the study, the men received medical treatment and insurance and other provisions to their families. When penicillin became the standard treatment for the disease in 1947, the medicine was not offered to any of the men in the study (neither the control nor experimental group). Additionally, the men were never informed about the life-threatening consequences of the treatments they were to receive and how they could infect their partners and children conceived once involved in the research. From 1932 to 1947, the date when penicillin was determined as a cure for the disease, dozens of men had died and numerous others had been infected.

An outcry led to an investigation of the study revealing that informed consent, a procedure where participants are fully informed about the research, its risks, and benefits and they provide voluntary and informed consent before participating, was not conducted. The participants were also not informed of the actual name of the study, "Tuskegee Study of Untreated Syphilis in the Negro Male," its purpose, and potential

Method	Description	Sources/Examples
Records/archival review	Reading archival information such as historical records, government documents, and archival maps provides historical and geopolitical context for the current environment.	Historical maps of roadways were used to understand how current lead soil contamination is correlated with the location of old highways (Rubio et al., 2022).
Observations	Observations allow us to understand people's interactions in the environment or context of interest. Given the sensitivity of the data of interest and the consent obtained, participants may or may not be aware that you are there to observe them.	
Interviews	One-on-one interviews (in person or virtual) offer a more personal way to collect data. Interviews are more time intensive than surveys but provide the opportunity to customize the questions for the interviewee and deviate from the script when interesting information arises. An interview can be formal and structured (where every interviewee receives the same questions) or informal and unstructured (more conversational and free flowing). A semi-structured interview is in between structured and unstructured where there is a planned list of questions but the interview is allowed to deviate from the list.	Post-huricane Katrina mobilized environmental and social justice movements. Read (X) to see how interviews were used to learn from community responses.
Focus groups	A focus group gathers a group of participants in a room (in person or virtual) where a discussion is moderated around the topic of concern. Rules for the discussion should be determined ahead of time (such as no interrupting). A focus group provides the opportunity for unique insights as participants will be influenced by each other. Focus groups are also the basis for additional participatory research methods including photovoice and participatory mapping, to name a few	
Survey	A questionnaire to capture information. The questions can cover open-ended, multiple choice, or fill in the blank. When designing your survey, consider who your target audience is and any eligibility and accessibility criteria required. A survey can provide both qualitative and quantitative information.	Learn about how citizen science projects surveyed people to build their advocacy strategies (Corburn, Jason, 2005).

TABLE 11.5   Summary of qualitative methods with reference examples of studies using the methods

consequences of the treatment they would receive during the study. T"e investigation also concluded that no choices were given to the participants to quit the study or"receive treatment when penicillin became available. The study was declared "ethically unjustified."

In response to abuses such as the Tuskegee syphilis study, the modern concept of the IRB was established to safeguard the rights, welfare, and well-being of individuals participating in research studies. The IRB is an independent committee that reviews and approves research protocols involving human subjects before they can be conducted. Its primary goal is to ensure that research involving humans is conducted ethically, with respect for participants' autonomy, privacy, and safety. IRBs are typically located at research universities but can also be externally operating or within a community, such as the Navajo Nation Human Research Review Board.

The importance of the IRB lies in its role as a critical safeguard for research participants. It helps ensure the following:

1. **Informed Consent:** Participants are fully informed about the research, its risks, and benefits, and they provide voluntary and informed consent before participating.

2. **Minimization of Risks:** Researchers are required to minimize risks to participants and ensure that potential benefits outweigh potential harms.

3. **Ethical Conduct:** Research is conducted with integrity and in accordance with ethical principles, respecting the dignity, autonomy, and privacy of participants. The location where data are collected, such as in a public setting, should account for the privacy of the participants.

4. **Privacy and Confidentiality:** Measures are in place to protect participants' personal information and maintain confidentiality, such as encrypting files and storing data in a secure location.

5. **Regular Oversight:** The IRB provides ongoing oversight of the research to ensure compliance with ethical standards and guidelines.

In summary, the IRB's history is rooted in the need to protect individuals from unethical research practices, and its continued importance lies in its role as a guardian of ethical conduct in research involving human subjects. It promotes the responsible and respectful treatment of participants, upholds the integrity of scientific research, and maintains public trust in the research enterprise.

The IRB process is important when working with community groups and should be discussed with stakeholders early in the process. Discuss with participants, particularly those who are providing data or a representative organization, what appropriate compensation for the data they provide is and what safeguards they require for the data collected.

Refer to the Collaborative Institutional Training Initiative (CITI Program) on Research, Ethics, and Compliance Training, particularly the training on human subjects research (see https://about.citiprogram.org for detailed information). Please review your institution's IRB requirements before proceeding.

11.2.2.7 Phase 2: Data Preprocessing and Data Splitting

Typical data science preprocessing steps include cleaning the data, normalizing, balancing a dataset, and assessing relevant features. Stakeholders can be decision makers in the data preprocessing steps through using surveys, interviews, and workshops to obtain stakeholder input on which features are important, relevant, and accurate. Stakeholders can also provide insight into how to address what to do with missing or duplicate data. For example, when you are trying to approximate missing data, you might consider dropping all of the missing data, or filling in the average value of the data, or using another statistical calculation that is appropriate. These decisions can be made in conversation with stakeholders, as they might give you insight into the data and how to deal with missing data points to avoid skewing the outcome of the model. Stakeholders can also help determine what portion of the dataset can be used for training and testing. For example, there could be geopolitical borders that split the dataset and might require the data to be relatively distributed from the regions so that they are representative of the geopolitical phenomena observed.

11.2.3   Phase 3: Model Development

During model development, including training the model, refining parameters, and determining appropriate metrics, the model can be presented back to stakeholders through many means, such as focus groups, workshops, websites, and videos. Important metrics of model performance (such as accuracy, recall, and F1 score) should be determined by the stakeholders. The data scientist is tasked with presenting the ML information in an accessible, inclusive, and clear format so that all stakeholders can understand the process and create an environment where stakeholders can engage with the process.

11.2.4   Phase 4: After Modeling

As a model becomes more complex, it can often turn into a "black box," making it harder to interpret and understand if the ML model predictions are consistent with fundamental principles of science (Zhong et al., 2021). It is necessary for the data scientists to ensure stakeholders are able to interpret the model. Contextualizing the data and maintaining the contextualization (such as who created the data, about which people does the data describe, and why were the data collected) are critical before applying the data to any decision-making algorithm.

Once a predictive model is created, stakeholders must be able to access, interpret, and dictate how the model is used. It is the responsibility of the data scientist to establish infrastructure to ensure data remain with vulnerable stakeholders and are not distributed without participant consent. To ensure this, data scientists have put together a Data Sheet checklist (see: https://arxiv.org/abs/1803.09010) to ensure that every ML dataset is accompanied with information about its source, operating characteristics, test results, and recommended uses (Gebru et al., 2018). We recommend following this checklist once you have developed your model.

11.2.4.1   Phase 4: Model Interpretation

Ideally, stakeholders should be able to interpret and understand the relationships within ML systems. Communities can often perceive ML as a mysterious black box that is uncertain and not guaranteed to work. Transparent ML systems, such as open source postings, do not necessarily imply improved understanding of the ML system without guiding context. Likewise, the availability of large "raw" ML datasets does not equate to transparency and does not promote public accessibility or meaningful public engagement with the data and results. Stakeholders can be engaged in the model interpretation through focus groups and stakeholder interviews, and it is the responsibility of the data scientist to inclusively and accessibly communicate the ML system and findings.

11.2.4.2   Phase 4: Model Applicability

The model needs to be validated against the context for which it will be applied. On the other hand, to prevent socio-technical gaps and unused systems, community partners must clearly understand what the ML system is capable of and its pitfalls. Data scientists and communities need to be honest from the project onset to ensure that community expectations of ML systems are realistic. For example, say data scientists want to design an ML system to automatically predict if an industrial company is violating water emissions regulations; however, in practice, the ML system may only identify indicators such as chemistry changes through water quality sensors. The system might require additional human efforts to verify whether the pollution event is indeed a violation and should be presented and stated as such.

11.2.4.3   Phase 4: Model Deployment

Model deployment means sharing and potentially allowing others to reuse or modify the model. Commonly used deployment approaches include sharing source code and providing executable files and web applications and can include workshops or forums where community partners are invited. Sharing source code for others to reuse requires expertise in coding; alternatively, web applications or executable files provide ready-to-use tools to make predictions but limit the ability of other researchers to modify or augment the tools (Zhong et al., 2021). Work with your community partner to determine how much they want to control and modify the model and help establish their capacity accordingly.

The relationship between the community and ML systems is a continuous adaptation process that spans long periods of time as communities are dynamic (Hsu et al., 2022). Likewise, ML systems need to adjust to the continuous changes in local communities. For instance, as we understand more about the real-life effects of ML systems on local people, we may need to fine-tune the underlying ML model using local community data. To do this, we may need to change and improve the data analysis process so that it fits the local community needs. We may even need to stop the ML algorithm under certain conditions. Such adaption requires ongoing commitment from researchers, designers, and developers to continuously maintain the infrastructure and build capacity for the community to be involved in the maintenance (Hsu et al., 2022).

The challenge remains of retaining long-term data scientist/community engagement with local people, especially in financially supporting local community members for their efforts. For example, community capacity needs to be built to manage and sustain the ML model, or funding is needed to hire software engineers who can maintain ML systems as community infrastructure in the long term.

11.3   How to Engage Communities in the Process: Case Study of the Mobile Lead Testing Unit Project in Newark, New Jersey

11.3.1   Brief History Newark Lead Crises

As we've discussed earlier in this chapter, lead is a heavy metal and prominent urban toxin that has caused, and continues to cause, significant harm in ESJ communities (Cassidy-Bushrow et al., 2017; Fedinick & Taylor, 2019; Ranganathan, 2016; Yeter et al., 2020). There is no safe level of lead in blood—even small concentrations can cause irreversible neurological and physiological damage (Cory-Slechta, 2012). Lead can enter the bloodstream through ingestion or inhalation.

Historical and modern discriminatory socioeconomic policy has resulted in a structurally racist infrastructure where low-income, Black, and other populations of color (collectively termed ESJ communities) are disproportionately affected by harmful toxins (Ranganathan, 2016). Due to this long-standing injustice, many communities are skeptical of politicians, institutions, and technological interventions. An EJ approach to childhood lead poisoning is a coordinated and comprehensive one that includes the creation of multilevel and multisectoral policies and partnerships. This includes the creation of jobs and job training; providing access to reliable transportation and food systems; providing safe, healthy, and affordable home ownership; and fostering ideas from those within and outside of low-income and minority neighborhoods to make a change (Whitehead & Buchanan, 2019).

In older cities, lead is found in multiple sources, including lead pipes, lead-based paint, soil, and dust. Newark, New Jersey, a city with a predominantly Black and immigrant population with a high poverty rate (~30 percent below the poverty line), is fraught with toxic industrial remnants, disproportionately exposing people of color, low income, and immigrant communities to toxin contamination (NRDC, n.d.). Newark, like other industrial cities, is revitalizing its aging, toxic infrastructure, and stakeholders have turned to data science as decision-making tools (for instance using ML algorithms to predict efficient lead service line replacement and identify areas where children are likely to have elevated blood lead levels). While retrofitting and rebuilding infrastructure, cities have an opportunity to reinvent decision-making processes to center equity. ML can be a tool through which multiple institutions and resources can be leveraged to reduce the social and environmental harms of toxic exposure while centering community voices in decision making. During the retrofitting process, we asked how we can design an ML algorithm such that the design process and outcomes are centering equity. What would justice look like in the context of Newark, and how can an ML algorithm work toward achieving it?

In 2016, Newark's drinking water garnered public attention with record-high lead levels (NRDC, n.d.). The city began handing out bottled water and point-of-use filters to homeowners, but later released a report detailing that without precise installation and maintenance, the filters do not prevent lead exposure, contributing to deep community skepticism of city-led interventions and decision makers. With pressure from residents, the EPA, and the National Resources Defense Council, Newark replaced all known lead service lines with copper pipes, leaving the premise plumbing as the only source of lead in drinking water. Like most old industrial cities, lead still remains in the paint and soil, which creates dust, leaving residents still vulnerable to exposure and harm.

11.3.2   Project Initiation: The Newark Water Coalition and the Initiation of the Mobile Lead Testing Unit

Spurred by the social and environmental injustices occurring in their community, the Newark Water Coalition, a grassroots community-based organization, was formed in response to the Newark Lead Crisis. Using a water filtration system, the organization distributes clean drinking to residents across Newark. As they evolved, the Newark Water Coalition aimed to look at holistic lead exposure from the environment inside

Newark residents' homes. The Newark Water Coalition is an anticapitalist, community-first organization driven to meet community needs of food, water, energy, and shelter. The Newark Water Coalition was primarily motivated by the poor drinking water quality in Newark but understands that lead poisoning, and health, are a product of multiple intersecting systems and must be addressed holistically.

Lead poisoning harms those most who experience, and are at the intersection of, class, economic, and racial oppression. The Newark Water Coalition advocates around these intersecting struggles in the community by distributing food and water to food-insecure residents; building food pantries; offering immigration, worker, and tenants' rights support; and providing policy advocacy. Climate change and gentrification are further exacerbating the inequitable infrastructure-induced stress on ESJ communities like those in Newark.

The Newark Water Coalition posed the research question, "Is lead poisoning happening in Newark? If so, where?" The Newark Water Coalition values emphasize collective thinking, shared responsibility, listening, taking into account the impact of current decisions on future generations, consensus decision making, and a holistic and intersectional view of health, all of which provide a strong premise for developing a community-driven research project. The UC Berkeley team is driven to extend the traditional science and engineering research practices and encourages scientists to immerse themselves in the field by taking on a social role and conducting hands-on field research. To support the Newark Water Coalition, the UC Berkeley team supplemented their research question with their own.

The Newark Water Coalition wanted to collect primary data from residents' homes measuring lead levels in water, soil, paint, and dust as well as a survey covering demographic and health information. The initiative was termed the Mobile Lead Testing Unit (MLTU). The MLTU will yield a dataset that can supplement ML models. The dataset and subsequent models will be one step toward helping water system managers and public health practitioners target resources to high-risk areas and allow the Newark Water Coalition to advocate for community science, lead remediation and abatement policies, and community support services and resources.

The MLTU Project was formed from a four-plus-year-long relationship between the Newark Water Coalition and the community and a two-year-long relationship between UC Berkeley and the Newark Water Coalition, both of which will continue on until the project is completed but has created a much longer connection. The collaboration between the Newark Water Coalition and UC Berkeley took two years to develop through UC Berkeley researchers having personal connection to New Jersey, traveling to New Jersey to join meetings, and assisting in food and water distributions. Following the key principles listed earlier, the project is centered on the community of Newark residents while recognizing there are sub-communities with individual identities. The relationship and understanding of the key functions and community services of the Newark Water Coalition allowed the data science project to be designed around the capacity, network, and resources of the Newark Water Coalition. The testing unit produced all of their advertising, informational materials, and surveys in multiple languages and canvased different locations throughout Newark.

When the Newark Water Coalition and UC Berkeley team started to design the study around an ML algorithm for infrastructure, they considered how the algorithm affects vulnerable populations at the intersection of oppressive structures.

11.3.3   Identifying Stakeholders

Stakeholders include decision makers who have control over pipe replacement, engineers who design infrastructure, advocacy groups, and grassroots organizations supporting residents affected by lead contamination. Using the knowledge of the Newark Water Coalition as well as interviews and community meetings, the stakeholders in Newark were mapped (Figure 11.3). The stakeholder map displays different government agencies, local organizations, and academic partners. Most stakeholders that we engaged with have some degree of power in the system but have varying levels of importance.

11.3.4   Data Collection

Once the relationship between the UC Berkeley team and the Newark Water Coalition team developed and the scope and timeline for the MLTU was decided, the UC Berkeley team joined the Newark Water Coalition team in Newark to commence the project. The MLTU team comprised Newark Water Coalition volunteers, all residents of Newark, from diverse backgrounds, ranging from high school students to high school teachers.

The testing unit relied on the strength of the Newark Water Coalition's relationship with community members, community leaders, and other organizations to recruit participants who trusted the team to enter their homes. The decisions on the study design, methods, recruitment, funding, development, and execution were done collaboratively between UC Berkeley and the Newark Water Coalition. Public workshops were held on community science, coding/data analysis, and building science education material to

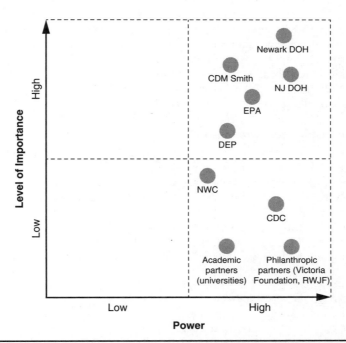

Figure 11.3   Map of Newark stakeholders covering lead exposure, remediation, and abatement.

Method	Data Collected	Details
Formal interviews	Remediation methods, existing datasets, public health resources	We interviewed stakeholders from local and state departments of health, industry companies working on lead pipe replacement, and grassroots organizations
Informal interviews	Best places for recruitment, how to design the MLTU house visits	Informal interviews occurred with stakeholders and residents of Newark over Zoom or in person at food distribution events
Focus groups	Resident concerns, perception, and experience with lead exposure and lead remediation/abatement processes	Small resident focus groups were held before the MLTU to inform the study design and survey questions
Qualitative survey	Demographic, socioeconomic, and health history information	The survey questions were co-written with the Newark Water Coalition and were informed by interviews with other stakeholders
Quantitative survey	Lead concentration measurements in paint, soil, dust, and water	In-field test kits used to measure lead concentration

TABLE 11.6   Data Collection Methods Used

help facilitate the integration and spread of information gained from the project. The project looked at health (exposure to lead) but was built upon an understanding that food, water, air, energy, and health are all interconnected. The main recruitment sites were at events that were centered on food, workers' rights, and tenants' rights events. Table 11.6 summarizes the different data collected by the MLTU.

Each resident received an individualized report with the findings from their home. The project is ongoing at the time of this textbook being published, but there are steps being taken to ensure the information gained in the study will be disseminated via reports, documentaries, printed materials, and events.

11.3.5   Building Community Capacity

Conducting a community-engaged project should have benefits for all parties involved. For data scientists, gaining community perspectives can make the outcome relevant for those affected, increase the rigor of the data collected because it contains the perspectives of those most harmed, and expand the reach of the results (Balazs & Morello-Frosch, 2013).

For the community, conducting research with data scientists can help them gain access to institutional resources (such as data processing systems and data science methods) and helps when defending the validity of their findings to other stakeholders.

11.4   Chapter Summary

This chapter introduces a community engagement process for designing an ML algorithm. The aim is to explain how communities can engage directly in gathering data for ML and how to use these data and the models for advocacy and activism. The case study in Newark, New Jersey, presents a close collaboration between stakeholders, particularly grassroots organizations and researchers, to build ML systems that can tackle local concerns with multiple benefits.

Chapter Glossary

Term	Definition
Community	In this chapter, we define a community as a group of people who are indirectly or directly affected by an issue and are working to ensure that these issues are recognized and resolved.
Community-based participatory research	CBPR is a collaborative approach to research that equitably involves all partners in the research process and recognizes the unique strengths that each brings. CBPR begins with a research topic of importance to the community and has the aim of combining knowledge with action and achieving social change to improve health outcomes and eliminate health disparities.
Data justice	Data justice means stakeholders, especially those marginalized by environmental and health impacts, have the possibility to understand, challenge, and be part of algorithmic decisions.
Environment justice	The Environmental Protection Agency defined environmental justice as "the fair treatment and meaningful involvement of all people regardless of race, color, national origin, or income with respect to the development, implementation, and enforcement of environmental laws, regulations, and policies."
Stakeholder	A stakeholder is any person, group, or organization with a vested interest or stake in the decision making and activities of a project, organization, or outcome.

References

Balazs, C. L., & Morello-Frosch, R. 2013. The three Rs: How community-based participatory research strengthens the rigor, relevance, and reach of science. Environmental Justice, 6 (1), 9–16. https://doi.org/10.1089/env.2012.0017

Benjamin, R. (2022). Viral justice: How we grow the world we want. Princeton University Press.

Billies, M., Francisco, V., Krueger, P., & Linville, D. (2010). Participatory action research: Our methodological roots. International Review of Qualitative Research, 3(3), 277–286. doi: 10.1525/irqr.2010.3.3.277

Cassidy-Bushrow, A. E., Sitarik, A. R., Havstad, S., Park, S. K., Bielak, L. F., Austin, C., Johnson, C. C., et al. (2017). Burden of higher lead exposure in african-americans starts in utero and persists into childhood. Environment International, 108, 221–227. https://doi.org/10.1016/J.ENVINT.2017.08.021

Clark, A. (2018). The Poisoned City: Flint's Water and the American Urban Tragedy. Metropolitan Books.

Cooper, C. B., et al. (2021). Inclusion in citizen science: The conundrum of rebranding: Does replacing the term 'citizen science' do more harm than good? Science, 372(6549), 1386. doi: 10.1126/science.abi6487

Corburn, J. 2005. Street Science: Community Knowledge and Environmental Health Justice. The MIT Press. https://doi.org/10.7551/mitpress/6494.001.0001

Cory-Slechta, D. A. (2012). Low Level Lead Exposure Harms Children: A Renewed Call for Primary Prevention (pp. 1–65).

Costanza-Chock, S. (2018). Design justice: Towards an intersectional feminist framework for design theory and practice. DRS2018 Catal., 2. doi: 10.21606/drs.2018.679

Costanza-Chock, S. (2020). Design pedagogies: 'there's something wrong with this system!'. Design Justice. doi: 10.7551/mitpress/12255.003.0009

Costanza-Chock, S. (2020). #TravelingWhileTrans, Design Justice, and Escape from the Matrix of Domination. Design Justice.

Dencik, L., & Sanchez-Monedero, J. (2022). Data justice. Internet Policy Review, 11(1), 1–16. doi: 10.14763/2022.1.1615

Dietrich, M., Shukle, J. T., Krekeler, M. P. S., Wood, L. R., & Filippelli, G. M. (2022). Using community science to better understand lead exposure risks. GeoHealth, 6(2). doi: 10.1029/2021GH000525

Drake, J. M. (2016, June 27). Expanding #ArchivesForBlackLives to traditional archival repositories. Retrieved September 25, 2017, from https://medium.com/on-archivy/expanding-archivesforblacklives-to-traditional-archival-repositories-b88641e2daf6

Fedinick, M. R., & Taylor, S. (2019). Watered Down Justice Report. National Resource Defense Council.

Flicker, S., Travers, R., Guta, A., McDonald, S., & Meagher, A. (2007). Ethical dilemmas in community-based participatory research: Recommendations for institutional review boards. Journal of Urban Health, 84(4), 478–493. doi: 10.1007/s11524-007-9165-7

Gebru, T., et al. (2018). Datasheets for Datasets. http://arxiv.org/abs/1803.09010

Holifield, R. (2001). Defining environmental justice and environmental racism. Urban Geography, 22(1), 78–90. doi: 10.2747/0272-3638.22.1.78

Holzmeyer, C. (2021). Beyond 'AI for Social Good' (AI4SG): Social transformations—not tech-fixes—for health equity. Interdisciplinary Science Review, 46(1–2), 94–125. doi: 10.1080/03080188.2020.1840221

Hsu, Y. C., Huang, T. H., Verma, H., Mauri, A., Nourbakhsh, I., & Bozzon, A. (2022). Empowering local communities using artificial intelligence. Patterns, 3(3). doi: 10.1016/j.patter.2022.100449

Irwin, A. (2001). Constructing the scientific citizen: Science and democracy in the biosciences. Public Understanding of Science, 10(1), 1–18. doi: 10.1088/0963-6625/10/1/301

Key, K. D., et al. (2019). The continuum of community engagement in research: A roadmap for understanding and assessing progress. Progress in Community Health Partnerships: Research, Education, and Action, 13(4), 427–434. doi: 10.1353/cpr.2019.0064

Kukutai, T., & Taylor, J. (2016). Indigenous data sovereignty: Toward an agenda (Vol. 38). ANU Press.

Lowry, C. S., & Stepenuck, K. F. (2021). Is citizen science dead? Environmental Science & Technology, 55(8), 4194–4196. doi: 10.1021/acs.est.0c07873

Montoya, M. J., & Kent, E. E. (2011). Dialogical action: Moving from community-based to community-driven participatory research. Qualitative Health Research, 21(7), 1000–1011. doi: 10.1177/1049732311403500

Morello-Frosch, R., Pastor Jr, M., Sadd, J. L., Porras, C., & Prichard, M. (2022). Citizens, Science, and Data Judo. In M. Minkler & N. Wallerstein (Eds.), Methods in Community-Based Participatory Research for Health (pp. 371). Wiley.

Morello-Frosch, R., Brown, P., Lyson, M., Cohen, A., & Krupa, K. 2011. Community voice, vision, and resilience in post-hurricane Katrina recovery. Environmental Justice, 4 (1), 71–80. https://doi.org/10.1089/env.2010.0029

NRDC. (n.d.). Newark Drinking Water Crisis. from https://www.nrdc.org/newark-drinking-water-crisis

O'Fallon, L. (n.d.). Successful Models of Community-Based Participatory Research. NIEHS.

Pace, C., Fencl, A., Baehner, L., Lukacs, H., Cushing, L. J., & Morello-Frosch, R. (2022). The drinking water tool: A community-driven data visualization tool for policy implementation. International Journal of Environmental Research and Public Health, 19(3), 1419.

Rainie, S. C., Schultz, J. L., Briggs, E., Riggs, P., & Palmanteer-Holder, N. L. (2017). Data as a strategic resource: Self-determination, governance, and the data challenge for Indigenous nations in the United States. The International Indigenous Policy Journal, 8(2).

Ranganathan, M. (2016). Thinking with flint: Racial liberalism and the roots of an American water tragedy. Capitalism Nature Socialism, 27(3), 17–33.

Rubio, J. M., et al. (2022). Use of historical mapping to understand sources of soil-lead contamination: Case study of Santa Ana, CA. Environmental Research, 212, 113478. doi: 10.1016/j.envres.2022.113478

Ulin, P. R., Robinson, E. T., & Tolley, E. E. 2004. "Qualitative Methods in Public Health: A Field Guide for Applied Research."

Vashney, K. (2022). Trustworthy Machine Learning. http://www.trustworthymachine-learning.com

Vera, L. A., Walker, D., Murphy, M., Mansfield, B., Siad, L. M., Ogden, J., & EDGI (2019). When data justice and environmental justice meet: Formulating a response to extractive logic through environmental data justice. Information, Communication & Society, 22(7), 1012–1028. doi: 10.1080/1369118X.2019.1596293

Wallerstein, N. B., & Duran, B. (2006). Using community-based participatory research to address health disparities. Health Promotion Practice, 7(3), 312–323. doi: 10.1177/1524839906289376

Wallerstein, N., & Duran, B. (2008). The theoretical, historical, and practice roots of community-based participatory research. In Minkler M. & Wallerstein N. (Eds.), Community-based participatory research for health: From process to outcomes (pp. 25–46).

Whitehead, L. S., & Buchanan, S. D. (2019). Childhood lead poisoning: A perpetual environmental justice issue? Journal of Public Health Management and Practice, 25(1), S115–S120. doi: 10.1097/PHH.0000000000000891

Williams, R. W. (1999). The contested terrain of environmental justice research: Community as unit of analysis. Social Science Journal, 36(2), 313–328. doi: 10.1016/s0362-3319(99)00008-7

Wilson, S., Campbell, D., Dalemarre, L., Fraser-Rahim, H., & Williams, E. M. (2014). A critical review of an authentic and transformative environmental justice and health community–university partnership. International Journal of Environmental Research and Public Health, 11(12), 12817–12834. doi: 10.3390/ijerph111212817

Yeter, D., Banks, E. C., & Aschner, M. (2020). Disparity in risk factor severity for early childhood blood lead among predominantly African-American Black children: The 1999 to 2010 US NHANES. International Journal of Environmental Research and Public Health, 17(5), 1552. doi: 10.3390/IJERPH17051552

Zhong, S., Zhang, K., Bagheri, M., Burken, J. G., Gu, A., Li, B., Ma, X. (2021). Machine learning: New ideas and tools in environmental science and engineering. Environmental Science & Technology, 55(19), 12741–12754.

End of Chapter Problems

1. Why are some communities more vulnerable to environmental health risks than others? Who is the most vulnerable?

2. Pick an environmental/health context, for example, a local design of stormwater management or agriculture discharge, and identify the stakeholders, community leaders, and decision makers.

3. Follow the steps in Section 11.2 to build a stakeholder map.

4. Reflect on your own positionality (race, class, gender, education, family, lived experiences, etc.), how you relate to different stakeholders, how you are perceived, and what resources your position allows you.

5. Choose a specific stakeholder in your system, and identify two to three questions you would ask them.

6. Look up your institution's institutional review board requirements. Complete the Collaborative Institutional Training Initiative (CITI Program) on Research, Ethics, and Compliance Training, particularly the training on human subjects research (https://about.citiprogram.org).

7. Find and evaluate an existing dataset following these questions from Datasheets for Datasets:

   a) For what purpose was the dataset created? Was there a specific task in mind? Was there a specific gap that needed to be filled? Please provide a description.

   b) Who created the dataset (e.g., which team, research group) and on behalf of which entity (e.g., company, institution, organization)?

   c) Who funded the creation of the dataset? If there is an associated grant, please provide the name of the grantor and the grant name and number.

8. Write a short response to the question from Viral Justice (Ruha Benjamin, 2022): Why is it important for scientific research initiatives to invite communities to be full partners in their work?

9. How have citizen scientists utilized their life experiences to advance research and investigation? How did citizen scientists fight for environmental justice in Flint, Michigan?

10. What steps can you take to ensure that institutions in your community, or institutions you work for, eliminate the practice of extractive logic in the name of science, research, or education?

11. What are the challenges of including community engagement in the data science design process, and how can you overcome these challenges?

12. What steps can you take and what methods can you use to share your ML findings with the community they impact and with the community from whom the data came?

Index

Note: Page numbers followed by *f* denote figures; by *t*, tables.

A

ABCDE for melanoma diagnosis, 141, 141*f*, 142
Abstract liberalism, 20
ACF plot. *See* Autocorrelation function (ACF) plot
ADDA. *See* Adversarial discriminative domain adaptation (ADDA)
Adversarial debiasing, 10–11, 13
Adversarial discriminative domain adaptation (ADDA), 76
Adversarial discriminator, 72
Adversarial methods, 72, 73, 79–81, 83
Adversarial training, 79–81
Aequitas, 11–12, 54, 55*t*, 56–57, 59
Agents of socialization, 19
Aggregation bias, 4
AI, Algorithmic, and Automation Incidents and Controversies Database, 94
AI Fairness 360. *See* AIF360
AI Incident Database, 94
AI journalistic robot, 95
AI/ML. *See* Machine learning (ML)
AIF360, 11, 54, 55*t*, 57–58
Algorithm, 20
Algorithmic bias, 18, 26
Algorithmic decision making, 138
Algorithmic fairness, 6–7, 148
Algorithmic fairness metric, 5
Algorithmic justice, 3–5
Algorithmic Justice League, 23, 31
Algorithms of Oppression (Noble), 24, 31, 106
Alibaba, 92
Allocative harm, 97
Amazon, 95

Anaconda, 127, 128
ANN. *See* Artificial neural network (ANN)
Anthropic, 92
Apple, 95
Archival review, 195*t*
"Are We Automating Racism" (2021), 24
ARIMA. *See* Autoregressive integrated moving average (ARIMA)
Artificial Intelligence Act (2021), 23
Artificial neural network (ANN), 141, 151
Aspirational equity, 160*f*, 161
ASR system. *See* Automated speech recognition (ASR) system
Association for Computing Machinery Code of Ethics, 21, 22
Attention layer, 112, 113
Attention layer mathematics, 112–113
Autism, 22
Autocorrelation function (ACF) plot, 39, 43*f*, 44*f*
Automated multiple pass method, 159
Automated speech recognition (ASR) systems, 95
Automating Inequality (Eubanks), 23
Autoregressive integrated moving average (ARIMA), 36
Average odds difference (AOD), 8, 13

B

Bard, 92
Behavioral bias, 98
Benchmark, 98
Benchmarking, 93
Benjamin, Ruha, 18

BERT, 77, 80, 81, 106, 110, 114, 120
Bias:
  aggregation, 4
  algorithmic, 18, 26
  behavioral, 98
  cognitive, 145, 148*t*
  defined, 13
  deployment, 5
  evaluation, 4, 146, 148*t*
  fairness toolkits, 50–51, 51*f*, 52*f*
  hate speech detection systems, 75
  health care, 139, 140*f*
  historical, 4, 97
  internet-trained models having internet-
    scale biases, 94
  label, 50, 51*f*, 97
  learning, 4, 98
  linking, 98
  measurement, 4
  mitigation of, ongoing process requiring
    constant adaptation, 85
  overamplification, 97
  representation, 4, 51, 52*f*, 97
  sampling. *See* Sampling bias
  selection, 97
  semantic, 97
  skin cancer, 142*f*
  socioecological inequity, 159
  sources of, 151
  statistical, 145, 146, 148*t*
  systematic error, 5
  technological, 2
  temporal, 98
  unconscious, 54
  underestimation, 147
  underrepresented communities, 50
Bias mitigation algorithm, 5, 9
Bidirectional encoder representations from
    transformers. *See* BERT
Bing, 92
"black" and "white," 150
Black box model, 93
Black swan theory, 129, 130
Black women, 35–36, 107
BLOOM, 93
Blue-black rule, 142
Bonilla-Silva, Eduardo, 20
BookCorpus, 94, 94*t*
Brockovich, Erin, 185
Buolamwini, Joy, 147

C

Calibration, 99
Causal reasoning, 51
CBPR. *See* Community-based participatory
    research (CBPR)
ChatGPT, 92, 93
CE survey. *See* Consumer Expenditure (CE)
    survey
Checklist, 53
CITI Program, 196
Citing sources, 24–25
Citizen science/community science, 184–186
Classification, 127
Classifier, 72
Claude, 92
Clustering, 159–160, 162–163, 169
Cognitive bias, 145, 148*t*
Colonization, 2
Commercial prediction algorithm, 134
Common Crawl, 94*t*, 106, 111, 114
Community-based participatory research
    (CBPR), 184, 185*t*
Community engagement for machine learning,
    177–207
  citizen science/community science, 184–186
  close collaboration among stakeholders,
    180, 182
  community-based participatory research
    (CBPR), 184, 185*t*
  community engagement, defined, 184
  data justice, 182–183
  environmental justice (EJ), 180, 181, 203
  Flint Water Crisis, 178–180
  glossary, 203
  lead pipes, 178–180, 198–202
  ML design. *See* ML design and community
    engagement
  Newark Lead Crises, 198–202
  what is a community?, 183–184, 203
Community science/citizen science, 184–186
Comorbidity, 130
Confusion matrix, 116
Consequentialism, 22–23
Consumer Expenditure (CE) survey, 163,
    164, 167
Contaminated examples, 3
Content categorization, 45
Copyright, 25–26
Correlation, 163
Counterfactual, 116

"Counterfactual Explanations Without Opening the Black Box" (Wachter et al.), 116
Counterfactual fairness, 7
Creative Commons, 25–26
Criminal recidivism, 50
Cultural expressions, 19
Cultural racism, 20
Curated dataset, 99

D

Data 4 Black Lives, 183
Data card, 115
Data centers, 93
Data collection, 97, 114–115, 193–196
Data curation tool, 115
Data documentation, 114, 115
Data equity framework, 161
Data generation, 3, 4*f*
Data inferential analysis, 39–44
Data justice, 182–183
Data lag, 40, 45
Data Nutrition Project, 115
Data privacy protections, 115
Data seasonality, 40, 45
Data Sheet checklist, 197
Data transparency, 114, 155
Dataset nutrition label, 114–115
"Dataset Nutrition Label: A Framework to Drive Higher Data Quality Standards, The" (Holland et al.), 115
DBSCAN, 163
DDI dataset. *See* Diverse Dermatology Images (DDI) dataset
Debiasing word embedding, 83
Decision tree (DT), 141, 151
DeepDerm, 144
Definitions. *See* Glossary
Demographic parity, 7, 116
Denigration, 98
Density-based spatial clustering, 163
Deontology, 22
Deployment bias, 5
Detroit Digital Justice Coalition, 183
Diagnostic metric, 98
Digital justice, 183
Digital storytelling, 24–26
Digital twins, 159
Discrimination, 19
Discrimination-aware data mining, 23

Disparate impact, 8, 14
Distorted sample, 3
Distributive equity, 160, 160*f*, 169
Diverse Dermatology Images (DDI) dataset, 143–145
Documentation debt, 114
DT. *See* Decision tree (DT)
DuBois, W. E. B., 161

E

EHE. *See* Environmental Hazard Evaluation (EHE)
EJ. *See* Environmental justice (EJ)
Environmental data justice, 183
Environmental Hazard Evaluation (EHE), 181
Environmental impact of using big data, 92–93
Environmental justice (EJ), 180, 181, 203
Environmental racism, 181
Equal opportunity, 7
Equal opportunity difference (EOD), 8, 14
Equal opportunity parity, 116
Equalized odds, 6–7
Equalized odds parity, 116
Equity:
    aspirational, 160*f*, 161
    defined, 3
    distributive, 160, 160*f*, 169
    exploitational, 160*f*, 161
    ostensible, 160*f*, 161
    procedural, 160, 160*f*, 169
    recognitional, 160–161, 160*f*, 169, 170
    systemic, 160. *See also* Systemic equity in socioecological systems
Equity-centered frameworks and tools, 170
ESJ communities, 199
Estimator bias, 146
Ethical and societal implications of machine learning, 17–32
    algorithmic bias, 18, 26
    Association for Computing Machinery Code of Ethics, 21, 22
    consequentialism, 22–23
    deontology, 22
    digital storytelling, 24–26
    discrimination-aware data mining, 23
    ethical imperatives, 22
    ethical implications of algorithms, 21–23
    fair machine learning, 23
    glossary, 26
    gut-check, 23

Ethical and societal implications of machine learning (*Cont.*):
  IEEE, 20
  informed consent, 21
  mitigating algorithmic bias, 23–24
  model card, 24
  Oath of Non-Harm for an Age of Big Data (Eubanks), 26, 27
  opacity, 22
  privacy, 21
  racial socialization, 19
  racism, 19–20
  societal/cultural implications of algorithms, 19–21
  systematic discrimination, 21
  virtue ethics, 23
Ethical imperatives, 22
Eubanks, Virginia, 23, 27
Euclidean space, 108
Europe:
  Artificial Intelligence Act (2021), 23
  General Data Protection Regulation (GDPR), 23, 115
Evaluation bias, 4, 146, 148*t*
Exnomination, 97
Explicit bias, 129–130
Exploitational equity, 160*f*, 161
Extrinsic bias metrics, 99

F

Facebook, 21
Facial recognition systems, 2, 50, 147, 159
Fair machine learning, 23
Fairlearn, 54, 55*t*, 57
Fairness metrics, 7–9
Fairness toolkits, 49–70
  Aequitas, 54, 55*t*, 56–57, 59
  AIF360, 54, 55*t*, 57–58
  bias, 50–51, 51*f*, 52*f*
  checklist, 53
  evaluating the toolkits, 55–56, 55*t*, 56–58
  Fairlearn, 54, 55*t*, 57
  fairness, 51–53
  future directions (future needs), 59
  glossary, 60
  limitations of toolkits, 58–59
  responsible AI, 53
  software toolkits, 53
Favorable label, 5, 14

First National People of Color Environmental Leadership Summit (1991), 181
Fitzpatrick skin typing test, 142, 143, 143*f*
"Fixing Medical Devices That Are Biased against Race or Gender" (Wallis), 30
Flagging, 117–118
Flint Water Crisis, 178–180
Focus group, 195*t*, 202*t*
Food spending, 163–167
Formal interview, 195*t*, 202*t*

G

GAN. *See* Generative adversarial network (GAN)
Gebru, Timnit, 147
Gender swap data augmentation, 83
General Data Protection Regulation (GDPR), 23, 115
General individual justice, 9
Generalized entropy index, 9
Generative adversarial network (GAN), 73
Glad You Asked (Raji), 24
Global nutrition transition, 170
Global Vectors for Word Representation, 107
Glossary:
  community engagement for machine learning, 203
  ethical and societal implications of machine learning, 26
  fairness toolkits, 60
  generally, 13–14
  hate speech detection systems, 86
  large language models (LLMs), 120–121
  medical ML/AI models, 134
  natural language processing, 100
  skin cancer, 151–152
  social media and health information, 45
  systemic equity in socioecological systems, 171
GLoVe, 107–108, 110, 114
Google:
  advanced search features, 26
  "Are We Automating Racism" (2021), 24
  automated speech recognition (ASR) systems, 95
  Bard, 92
  categorizing African Americans as "gorillas," 159
  criminal background checks/mugshot viewing sites, 50

Google (*Cont.*):
   ethical implications of algorithms, 21
   search algorithm, 107
"Google apologises for Photos app's racist
   blunder" (2015), 159
GPT, 106
GPT-1, 93
GPT-2, 93, 110
GPT-3, 93, 110, 114
GPT-3.5, 93
GPT-4, 92
Group fairness, 6, 14
Guidelines for Trustworthy AI (2020), 99
Gut-check, 23
Gutenberg Corpus, 94, 94*t*

H

HAM10000, 143, 144
Hate speech detection systems, 71–90
   adversarial methods, 72, 73, 79–81, 83
   BERT model, 77, 80, 81
   bias, 75
   case study, 73–74
   challenges, 73
   complex and multifaceted task requiring
      diverse array of skills, 86
   glossary, 86
   hands-on exercise, 83–85
   long short-term memory (LSTM), 73, 85
   multitask learning, 72, 73, 77–79, 81–82, 85
   other methods of mitigating bias, 82–83
   random forest, 84
   recurrent neural network (RNN), 73
   transfer learning, 72, 73, 75–77, 81
   what are they/how they work?, 72–73
H.CUP. *See* Healthcare Cost & Utilization
   Project (H.CUP)
Health care:
   bias, 139, 140*f*
   diagnostic ML tools, 138
   "Fixing Medical Devices That Are Biased
      against Race or Gender" (Wallis), 30
   labels ("black" and "white"), 150
   medical ML systems. *See* Medical ML/AI
      models
   melanoma. *See* Skin cancer
   mental health, 34–35, 36, 130, 131*f*, 132*f*, 132*t*
   social media. *See* Social media and health
      information
Health Inequality Project, 130

Healthcare Cost & Utilization Project
   (H.CUP), 131
Heart attack, 22
Hierarchical clustering, 163
Historical bias, 4, 97
HIV. *See* MyHealthImpactNetwork
HuggingFace, 92
Human immunodeficiency virus (HIV).
   See MyHealthImpactNetwork

I

IBM, 95
IBM's AI Fairness 360. *See* AIF360
IDE. *See* Integrated development environment
   (IDE)
IEEE. *See* Institute for Electronics and Electrical
   Engineering Code of Ethics (IEEE)
ImageNet database, 147
Implicit bias, 130
In-processing, 9–10, 14
Indigenous Data Sovereignty movement, 183
Individual fairness, 6, 14
Informal interview, 195*t*, 202*t*
Informed consent, 21
Institute for Electronics and Electrical
   Engineering Code of Ethics (IEEE), 20
Institutional review board (IRB), 190, 194, 196
Integrated development environment (IDE), 127
Interviews, 195*t*, 202*t*
Intrinsic bias metrics, 98–99
ipython, 128
IRB. *See* Institutional review board (IRB)
ISIC, 143

J

jupyter-matplotlib visualization library, 128
Jupyter Notebook, 128
Justice 40 initiative, 181

K

k-means clustering, 163
k-nearest neighbor algorithm, 129
Kaiser Family Foundation, 44
Key vector, 113
King, Martin Luther, Jr., 106
KPSS. *See* Kwiatkowski-Phillips-Schmidt-Shin
   (KPSS) test
Kwiatkowski-Phillips-Schmidt-Shin (KPSS)
   test, 43

L

Label bias, 50, 51*f*, 97
Labels ("black" and "white"), 150
"Lack of Transparency and Potential Bias in Artificial Intelligence Data Sets and Algorithms" (Daneshjou et al.), 143
Large language models (LLMs), 105–124
 attention layer mathematics, 112–113
 bad data in, bad data out, 106–107
 counterfactuals, 116
 data card, 115
 data collection, 114–115
 data documentation, 114, 115
 dataset nutrition label, 114–115
 existing web resources, 94
 fighting bad math with better math, 117–119
 flagging, 117–118
 glossary, 120–121
 GLoVe, 107–108, 110
 linear analogies, 108–110
 linear decision making for nonlinear language, 111–112
 linear mapping, 111
 matrix factorization, 108
 Microsoft's Tay, 119–120
 model constraints/operations, 117–118
 nudging, 118
 parity, 116
 probability of co-occurrence, 107
 pruning, 118
 stratified sampling, 117
 transformer, 113, 114
 vectorization, 108
 word embedding, 111, 114, 121
Lead pipes, 178–180, 198–202
Learning bias, 4, 98
LeMoult, Craig, 30
Lesion segmentation, 141, 142
LIME. *See* Local interpretable model-agnostic explanations (LIME)
Linear analogies, 108–110
Linear mapping, 111
Linear regression analysis, 45
Linking bias, 98
Little Mix, 95
LLMs. *See* Large language models (LLMs)
Local interpretable model-agnostic explanations (LIME), 58
Logistic regression, 131

Long short-term memory (LSTM), 73, 85
LSTM. *See* Long short-term memory (LSTM)
Lung cancer, 133*f*

M

Machine learning (ML):
 AI and ML use in socioecological systems, 158–159
 algorithmic decision making, 138
 best practices to build fairer application, 12, 12*f*
 bias, 7
 causes of injustice, 3
 clustering, 159–160
 defined, 18, 126
 ethics. *See* Ethical and societal implications of machine learning
 fairness (algorithmic fairness), 148
 historical mentions of "machine learning" in published books, 126*f*
 limitations of ML models, 147*t*
 methods for achieving fairness, 9–11
 NLP. *See* Natural language processing (NLP)
 origin of the term "machine learning," 126
 overcoming traditional statistical limitations, 186
 population data feedback loop, 139*f*
 social and environmental applications, 186–187, 187*t*
 social media and health information, 44–45
 sources of harm in ML life cycle, 3–5
 supervised/unsupervised learning, 126, 127
 unfairness in ML decision making, 145
Manning, Christopher, 107
MATLAB, 163–166
MATLAB-PCA clustering, 166, 167*f*
MATLAB-PCA scattering, 164*f*, 165
MATLAB scattering, 164–165, 164*f*
Matrix factorization, 108
Mean square error (MSE), 146
Measurement bias, 4
Medical ML/AI models, 125–136
 black swan theory, 129, 130
 comorbidity, 130
 glossary, 134
 Health Inequality Project, 130
 Healthcare Cost & Utilization Project (H.CUP), 131
 lung cancer, 133*f*

Medical ML/AI models (*Cont.*):
  overfitting/underfitting, 129
  patient length of stay, 133*f*
  severe mental illness (SMI), 130, 131*f*,
      132*f*, 132*t*
  systematic error, 134
  use cases, 127–134
Mental health, 34–35, 36, 130, 131*f*, 132*f*, 132*t*
MICE. *See* Multivariate Imputation by Chained
    Equations (MICE)
Microsoft, 92, 95, 119–120
Microsoft's Fairlearn. *See* Fairlearn
Minimization of racism, 20
ML. *See* Machine learning (ML)
ML design and community engagement,
    187–198
  building partnerships, 188
  data collection, 193–196
  data preprocessing and data splitting,
      196–197
  Data Sheet checklist, 197
  formulating the question, 190
  institutional review board (IRB), 190,
      194, 196
  model development, 197
  model interpretation, applicability, and
      deployment, 197–198
  overview, 189*f*
  stakeholder importance map, 191*f*, 192
  stakeholder matrix, 192, 193*t*
  stakeholders and stakeholder relationships,
      190–193
MLFlow, 115
Model building and implementation, 4–5, 5*f*
Model card, 24
Model parity, 116
ModelDerm, 144
MSE. *See* Mean square error (MSE)
Multicollinearity, 38–39
Multitask learning, 72, 73, 77–79, 81–82, 85
Multivariate Imputation by Chained Equations
    (MICE), 38
Musk, Elon, 34, 92
MyHealthImpactNetwork, 35–39
  Black women, 35–36
  CorHeatMap of continuous variables, 39, 40*f*
  data preprocessing, 36
  imputation diagnosis using density plot,
      38, 39*f*
  key performance indicators (KPIs), 34

MyHealthImpactNetwork (*Cont.*):
  landing page and Twitter feed, 35*f*
  missing values, 37–38, 38*f*
  multicollinearity, 38–39
  online HIV prevention awareness
      platform, 35
  variables used for analyses, 36, 37*t*

N

National Health and Nutrition Examination
    Survey (NHANES), 159
National Public Radio (NPR), 25
Natural language generation (NLG), 99
Natural language processing (NLP), 91–104
  allocative harm, 97
  Amazon's internal recruiting tool, 95
  automated speech recognition (ASR)
      systems, 95
  bias classification, 97–98, 97*f*
  challenges for mitigating biases, 93
  data collection, 97
  defined, 45
  denied opportunities and preconceived
      views, 96–97
  enabling computer program to understand
      human language, 44
  environmental impact of using big data,
      92–93
  exnomination, 97
  glossary, 100
  internet-trained models having internet-
      scale biases, 94
  intrinsic/extrinsic bias metrics, 98–99
  large language models (LLMs), 94
  mitigating NLP bias and unfairness, 98–99
  natural language generation (NLG), 99
  natural language understanding (NLU), 99
  new releases of generative models, 92
  preprocessing and feature extraction
      techniques, 99
  reported incidents of discrimination,
      94–95
  representational harm, 96
  risks, 92
  selecting photo of wrong mixed-race
      person, 95
  sociocognitve fallacies, 93
  sociocognitve taxonomy, 98
  subfields of NLP, 99
  trusting AI, 93

Natural language processing (NLP) (*Cont.*):
using general measurement processes to
examine NLP biases, 99
Natural language understanding (NLU), 99
Naturalization, 20
Navajo Nation Human Research Review
Board, 196
Negative correlation, 163
Newark Lead Crises, 198–202
NHANES. *See* National Health and Nutrition
Examination Survey (NHANES)
NLG. *See* Natural language generation (NLG)
NLP. *See* Natural language processing (NLP)
NLU. *See* Natural language understanding
(NLU)
Noble, Safiya, 24, 31, 106
Non-neutrality, 2
NPR. *See* National Public Radio (NPR)
Nudging, 118

═══ **O** ═══

Oath of Non-Harm for an Age of Big Data
(Eubanks), 26, 27
Observations, 195*t*
Office of Environmental Justice, 181
Opacity, 22
Open source postings, 197
OpenAI, 92
Ostensible equity, 160*f*, 161
Overamplification bias, 97
Overfitting, 129, 193–194

═══ **P** ═══

PACF plot. *See* Partial autocorrelation function
(PACF) plot
Pacific Gas & Electric, 185
Parity, 116
Partial autocorrelation function (PACF) plot,
39, 43*f*, 44*f*
Participatory action research, 184
Patient length of stay, 133*f*
"Pause Giant AI Experiment," 92
PCA. *See* Principal component analysis (PCA)
PEAR Plan & Playbook, 161
Pearson correlation test, 39, 40*f*
Peck, Evan, 29
Pennington, Jeffrey, 107
Perceptions, 145–146
Perceptron equation, 146, 146*f*
Perceptron function, 146

Permission-to-pollute system, 181
Perturbation and counterfactuals, 99
PH2, 143
Pinnock, Leigh-Anne, 95
Population data feedback loop, 139*f*
Positive correlation, 163
Postprocessing, 9–10, 14, 22
Predictive policing, 21
Prejudice, 19
Preprocessing, 9
Principal component analysis (PCA), 163
Privacy, 21
Pro-Equity Anti-Racism (PEAR) Plan &
Playbook, 161
Probability of co-occurrence, 107
Procedural equity, 160, 160*f*, 169
Project Implicit, 29
Prospector static code analysis tool, 128
Protected attribute, 5, 14
Protected variable, 10
Pruning, 118
Public domain materials, 26
Python, 128, 163, 166

═══ **Q** ═══

Qualitative data, 194
Qualitative research, 194
Qualitative survey, 202*t*
Quantitative survey, 202*t*
Query vector, 113

═══ **R** ═══

Race After Technology (Benjamin), 18, 114
Racial socialization, 19
Racialized social system, 20
Racism, 19–20
Raji, Deborah, 24
Random forest, 84, 131
Recognition, 98
Recognitional equity, 160–161, 160*f*, 169, 170
Records/archival review, 195*t*
Recurrent neural network (RNN), 73
Reddit, 111
Reddit Corpus, 94, 94*t*
Regression, 127
Regularization, 147
Reinforcement learning, 118
Reject option-based classification, 11, 14
Representation bias, 4, 51, 52*f*, 97
Representational harm, 96

Responsible AI, 53
Reweighing, 10, 14
RNN. *See* Recurrent neural network (RNN)
RoBERTa, 106, 110, 114

━━━ **S** ━━━

Sample size disparity, 3
Sampling bias
    bias classification, 97*f*
    nonrandom sampling of the population, 97
    seemingly unproblematic data gathering, 22
    skewed results, 51, 52*f*
    skin cancer, 142, 146, 147, 148*t*
Scatter plot visualization, 164–167
Scientific citizens, 188
Scikit-Learn, 129
Selection bias, 97
Self-attention layer, 113
Semantic bias, 97
Semi-structured interview, 195*t*
Sensitive attribute, 51
Sensitive information, 59
Sentiment analysis, 45
Severe mental illness (SMI), 130, 131*f*,
    132*f*, 132*t*
Similarity measures, 51
Skin cancer, 137–156
    ABCDE for melanoma diagnosis, 141,
        141*f*, 142
    AI-generated images of melanoma, 140*f*
    bias, 142*f*
    blue-black rule, 142
    case study (mitigating bias in ML for
        melanoma), 139–145
    challenges arising when detecting cutaneous
        malignancies, 144
    cognitive bias, 145, 148*t*
    computer-generated images of melanoma
        lesions, 150, 150*f*
    computer vision algorithms, 141
    DDI dataset, 143–145
    evaluation bias, 146, 148*t*
    Fitzpatrick skin typing test, 142, 143, 143*f*
    glossary, 151–152
    lesion segmentation, 141, 142
    light-skinned/dark-skinned populations, 140*f*
    mitigation techniques, 148*t*, 149, 149*f*
    processes for ML melanoma detection,
        141, 142
    regularization, 147

Skin cancer (*Cont.*):
    retraining detection algorithms for diverse
        skin tones, 142–143
    sampling bias, 146, 148*t*
    stages of melanoma and survival rates, 143*f*
    statistical bias, 145, 146, 148*t*
    underestimation bias, 147
    visual consensus label performance for
        training AI models, 144
SMI. *See* Severe mental illness (SMI)
Socher, Richard, 107
Social media and health information, 33–48
    ACF and PACF plots, 39, 43*f*, 44*f*
    Black women, 35
    daily total potential reach, 39, 41*f*, 42*f*
    data inferential analysis, 39–44
    ethical issues, 44
    glossary, 45
    key performance indicators (KPIs), 34, 35
    lags, 40, 45
    machine learning, 44–45
    mental health, 34–35, 36
    mitigating bias, 34–35
    scholarly health research network. *See*
        MyHealthImpactNetwork
    seasonality, 40, 45
    time series analysis, 39–44
    Twitter, 34, 35, 35*f*, 36
Socialization, 19
Societies, 19
Sociocognitve fallacies, 93
Socioecological systems, 169–170. *See also*
    Systemic equity in socioecological
    systems
Software toolkits, 53
"Some Studies in Machine Learning Using the
    Game of Checkers" (Samuel), 126
Spärck Jones, Karen, 111
Stakeholder importance map, 191*f*, 192
Stakeholder matrix, 192, 193*t*
Statistical bias, 145, 146, 148*t*
Statistical measures, 51
Statistical parity, 6, 7–8
Statistical parity difference (SPD), 8, 14
Statistics of Common Monthly Archive, 114
Stereotyping, 98
Stochastically robust process, 112
Stochasticity, 112
Stratified sampling, 117
Structural racism, 20

Structured interview, 195*t*
Subsidies, 167
Supervised machine learning, 126, 127
Support vector machine (SVM), 45, 141, 151
Survey, 195*t*, 202*t*
SVM. *See* Support vector machine (SVM)
Systematic discrimination, 21
Systematic error, 134
Systemic equity in socioecological systems, 157–176
   basic concepts and definitions, 160
   clustering, 162–163, 169
   examples of socioecological inequity and bias, 159
   food spending, 163–167
   global nutrition transition, 170
   glossary, 171
   MATLAB-PCA clustering, 166, 167*f*
   MATLAB-PCA scattering, 164*f*, 165
   MATLAB scattering, 164–165, 164*f*
   nexus of food, energy, and water, 169
   PEAR Plan & Playbook, 161
   scatter plot visualization, 164–167
   socioecological systems, 169–170
   systemic equity framework, 160–161, 160*f*
   Wells-DuBois protocol, 161, 162*t*, 167, 168*t*, 170, 171

T

Tay, 119–120
Technological bias, 2
Temporal bias, 98
Terminology. *See* Glossary
TF-IDF. *See* Time frequency-inverse document frequency (TF-IDF)
Theil index, 9, 14
Thirwall, Jade, 95
TikTok, 26
Time frequency-inverse document frequency (TF-IDF), 84
Time series analysis, 39–44
Token, 106
Tongyi Qianwen, 92
"Toxic Wastes and Race in the United States," 181
Transfer learning, 72, 73, 75–77, 81, 144, 145
Transformer, 113, 114
Transparent ML system, 197
True positive rate, 8

"Tuskegee Study of Untreated Syphilis in the Negro Male," 194
Tuskegee syphilis study, 194, 196
Twitter, 34, 35, 35*f*, 36, 111. *See also* Social media and health information

U

UChicago's Aequitas. *See* Aequitas
Unconscious bias, 54
Underestimation bias, 147
Underfitting, 129
Underrepresentation, 98
Underrepresented minorities (URM) population, 130
United States Indigenous Data Sovereignty Network, 183
Unstructured interview, 195*t*
Unsupervised machine learning, 126, 127
URM. *See* Underrepresented minorities (URM) population
User error, 18

V

Value vector, 113
Vector, 106
Vectorization, 108
Virtue ethics, 23

W

Wells, Ida B., 161
Wells-DuBois protocol, 161, 162*t*, 167, 168*t*, 170, 171
White box model, 93
#WhyWeTweetMH, 36
Wikipedia, 94, 94*t*, 111
Word embedding, 98, 106
Wozniak, Steve, 92

X

X, 34. *See also* Twitter
XGboost, 131, 133*f*

Y

YouTube, 21

Z

Z stereotyping, 10